Engineering the Circular Economy

Implementing the circular economy necessitates innovative approaches capable of managing the complexity and uncertainty inherent in the shift from linear to circular systems. Fuzzy logic, a computational approach inspired by human reasoning, provides a promising framework for navigating these complexities. With a practical approach, this book offers how fuzzy logic can be utilized to implement the circular economy.

Engineering the Circular Economy: Leveraging Fuzzy Logic for Implementation discusses the significance of incorporating fuzzy logic into circular economy strategies. This integration enables the development of adaptive models for resource management, waste reduction, and supply chain optimization. The book showcases various case studies and practical examples to demonstrate the application of fuzzy logic in implementing circular economy principles across diverse industries. Additionally, it addresses the challenges and future directions within a circular economy framework, emphasizing the importance of identifying these challenges and exploring future directions for continuous improvement in sustainability efforts. The potential of combining engineering, fuzzy logic, and the circular economy to address global sustainability challenges is highlighted, showcasing how leveraging fuzzy logic facilitates informed decision-making and the creation of innovative solutions as we work toward a more sustainable future.

With its forward-thinking approach and focus on practical applications, this book is an essential resource for practitioners, researchers, engineers, scholars, and policymakers looking to promote innovation and sustainability in today's rapidly changing economic and environmentally conscious environment.

Sustainable Industrial Engineering and Management Systems

Engineering the Circular Economy: Leveraging Fuzzy Logic for Implementation
Biswadip Basu Mallik, Gunjan Mukherjee, Rahul Kar, and Youqing Fan

Engineering the Circular Economy

Leveraging Fuzzy Logic for Implementation

Edited by

Biswadip Basu Mallik, Gunjan Mukherjee, Rahul Kar, and Youqing Fan

CRC Press
Taylor & Francis Group
Boca Raton London New York

CRC Press is an imprint of the
Taylor & Francis Group, an **informa** business

Designed cover image: Shutterstock ©

First edition published 2026
by CRC Press
2385 NW Executive Center Drive, Suite 320, Boca Raton FL 33431

and by CRC Press
4 Park Square, Milton Park, Abingdon, Oxon, OX14 4RN

CRC Press is an imprint of Taylor & Francis Group, LLC

ISBN: 978-1-032-88560-5 (hbk)
ISBN: 978-1-032-88639-8 (pbk)
ISBN: 978-1-003-53883-7 (ebk)

DOI: 10.1201/9781003538837

Typeset in Times
by Apex CoVantage, LLC

Contents

Preface

The dawn of the 21st century has brought with it a renewed focus on sustainability and responsible resource management. In the face of mounting environmental challenges such as climate change, pollution, and resource depletion, the concept of the circular economy has emerged as a beacon of hope for achieving a more sustainable future. At its core, the circular economy is a holistic approach to economic development that aims to redefine growth, focusing on maximizing the value of resources and minimizing waste. Overall, the new trend of such technology has played a crucial role in the optimization of the overall behavior. Unlike the traditional linear model of production and consumption, which follows a "take–make–dispose" pattern, the circular economy seeks to close the loop, ensuring that materials and resources are reused, recycled, or regenerated at the end of their life cycles. Engineering the circular economy requires a multidisciplinary approach, combining principles from engineering, economics, environmental science, and social sciences. It demands innovative solutions, transformative technologies, and novel strategies for redesigning products, processes, and systems to minimize waste and maximize resource efficiency. In this context, the application of fuzzy logic emerges as a powerful tool for navigating complexities inherent in the transition to a circular economy. Fuzzy logic, a branch of artificial intelligence that deals with reasoning and decision-making in the presence of uncertainty and imprecision, offers a flexible framework for modeling and optimizing complex systems with vague or incomplete information. By leveraging fuzzy logic, engineers and policymakers can develop robust decision-support systems, predictive models, and optimization algorithms tailored to the unique challenges of implementing circular economy initiatives. Whether in the design of closed-loop supply chains, the optimization of waste management processes, or the development of sustainable product-service systems, fuzzy logic provides a means to handle the inherent ambiguity and uncertainty of real-world scenarios. The economic uncertainties and its optimizing effects can well be measured and monitored by the very in-depth concept of the fuzzy logic approach.

This book explores the intersection of engineering and fuzzy logic in the context of the circular economy, offering insights, case studies, and practical strategies for fostering sustainable development. Through a blend of theoretical discussions and practical applications, it aims to equip readers with the knowledge and tools needed to engineer innovative solutions that promote resource efficiency, environmental stewardship, and economic prosperity. The close-knit connection among the multiple allied disciplines has been exposed in some integrated fashion in this book. As we stand at the precipice of a critical juncture in human history, the imperative to transition to a circular economy has never been more urgent. By harnessing the power of fuzzy logic and embracing the principles of sustainability, we can pave the way toward a future where economic prosperity is decoupled from environmental degradation, and where the well-being of both present and future generations is safeguarded.

This book is the short anecdote of the research journey toward engineering a circular economy—one fuzzy logic decision at a time. The presentation followed in this book is very lucid and easy to understand for all categories of readers. The presentation of the topics and its concerned exercise is very worthwhile and appropriate to the readers of diverse fields. Overall, the book provides a wholesome approach to the circular economies in terms of fuzzy-based analysis in pretty significant manner.

Dr. Biswadip Basu Mallik
Dr. Gunjan Mukherjee
Rahul Kar
Dr. Youqing Fan
Editors

About the Editors

Biswadip Basu Mallik, PhD, is a professor of Mathematics in the Department of Basic Science & Humanities at the Institute of Engineering & Management (School of University of Engineering and Management), Kolkata, India. With 23 years of experience in teaching and research, he has published numerous research papers and book chapters in scientific journals from esteemed international publishers such as CRC Press, Apple, Wiley, IGI Global, Springer, Nova Science, IEEE, etc. Dr. Basu Mallik has authored five undergraduate-level books on engineering mathematics, quantitative methods, and computational intelligence. He has also published five Indian patents and edited twenty-one books. His research interests include computational fluid dynamics, mathematical modeling, optimization, and machine learning. He serves as the series editor for the book series titled Mathematics and Computer Science from Wiley-Scrivener, USA. He is the managing editor of the *Journal of Mathematical Sciences & Computational Mathematics* (JMSCM), USA. Additionally, Dr. Basu Mallik is editorial board member and reviewer for several scientific journals. He holds senior life membership in the Operational Research Society of India (ORSI) and life membership in the Calcutta Mathematical Society (CMS), Indian Statistical Institute (ISI), Indian Science Congress Association (ISCA), and the International Association of Engineers (IAENG).

Gunjan Mukherjee, PhD, is an associate professor in the Department of Computational Sciences, Brainware University, Kolkata, West Bengal, India. He has completed his BSc in physics (CU), MCA (IGNOU), MTech (CU) and obtained PhD degree in engineering from Jadavpur University, Kolkata, West Bengal, India. His research interest includes computer vision, machine learning, soft computing, image processing, etc. He is the life member of CSI, ISOC, and IAENG. He has published a number of papers in different journals and international conferences of repute. Dr. Mukherjee also guided many students of BTech and MCA in their project and research works. He also authored the school computer book series (class 3–10), a textbook on IT systems theory for engineering students, and a book on python technology (in press). He has also been the reviewer of many technical books and an editor of a number of books. Dr. Mukherjee has experience in different sectors of teaching for nearly 20 years.

Rahul Kar is a faculty member of Mathematics at Kalyani Mahavidyalaya, Nadia, West Bengal, India. He has completed his MSc in Pure Mathematics and pursuing PhD in Mathematics from Bankura University, West Bengal, India. He has worked as a guest faculty of Gurunanak Institute of Technology and other two colleges. In total, he has eight years of teaching experience. He has published many research papers with Scopus index. He is also working as an editor of four national and international journals and a reviewer of five international journals. He is working as an editor of

15 mathematics/computer books. He is an annual member of Indian Mathematical Society as well as honorary member of Carmels Research Institute, Dakar, Senegal.

Youqing Fan, PhD, is an associate professor and chief investigator of an ARC Linkage Project on how volunteering can be better managed to become more accessible to CaLD elderly in Australia ($331,000) and a School-funded PRI project ($75,000). He is associate editor for *Journal of Business Research* (ABDC A; Scimago Q1), and *Employee Relations* (ABDC B; Scimago Q1); and an inaugural A/editor for *Journal of Tropical Futures*. Youqing is the recipient of the 2023 School Citation for Research Award. His research and teaching focuses on the interaction of organizational management and sustainability, with a specific interest in understanding how managers incorporate the environmental, social, and economic aspects of sustainability goals into their organizational management practices at the workplace and along the supply chain. Dr. Fan has contributed significantly to the fields of sustainability and business management. He has published 20 papers in ABDC A* or A or Scopus Q1 journals, including papers published in major business journals such as the *Journal of Business Research, International Journal of Human Resource Management, Business Strategy and the Environment, Enterprise Information Systems, Cross-Cultural and Strategic Management, Journal of Global Information Management*, and *Asian Business and Management*, and in high-impact sustainability journals such as *Transportation Research Part E: Logistics and Transportation Review, Journal of Environmental Management, Journal of Cleaner Production, Corporate Social Responsibility & Environmental Management and Habitat International.* He has published an edited book on addressing employee mental health issues, along with several *The Conversation* articles. He is also a regular reviewer for *Journal of Business Research, International Journal of Human Resource Management, International Journal of Hospitality Management, Academy of Management Annual Conference, International Journal of Production Research, Information and Management, Sustainable Development*, and *Economic and Labour Relations Review*. Dr. Fan's research findings have generated social impacts beyond academia. His viewpoints appear in independent academic media such as TheConversation.com and The Fifth Estate, and the mainstream media outlets such as ABC News and Sydney Morning Herald. Dr. Fan has successfully led a School funded Priority Research Initiative Project to completion. The project examines urban employment and housing segregation. Dr. Fan has also been involved in international collaborative projects investigating the poverty reduction and migrant health inequalities in the global south. Dr. Fan has successfully supervised three PhD students to completion. As of April 2023, he has published four journal papers with PhD students.

Contributors

S. Balakrishnan
Aarupadai Veedu Institute of Technology
Vinayaka Mission's Research Foundation (Deemed to be University)
Chennai, Tamil Nadu, India

P. Balasubramani
Department of Computer Science and Engineering
KIT-Kalaignarkarunanidhi Institute of Technology
Kannampalayam, Tamil Nadu, India

Sohini Banerjee
Department of Computer Science
Abacus Institute of Engineering and Management
Mogra, West Bengal, India

Biswadip Basu Mallik
Department of Basic Science and Humanities
Institute of Engineering & Management
(School of University of Engineering and Management)
Kolkata, West Bengal, India

Aditi Bhardwaj
Department of Computer Science Engineering
ShriRam Institute of Information Technology
Morena, Madhya Pradesh, India

Basit Ali Bhat
Gulzar Group of Institutes
Ludhiana, Punjab, India

Sandip Chakraborty
Department of Computational Sciences
Brainware University
Kolkata, West Bengal, India

A. Christo
Faculty of Intelligent Systems and Computer Technologies
Samarkand State University
Samarkand, Uzbekistan

Sudeshna Das
Department of Computer Science
Abacus Institute of Engineering and Management
Mogra, West Bengal, India

Maulika Premal Desai
Department of Commerce
Christ University
Bengaluru, Karnataka, India

Toijam Sarika Devi
Centre For Management Studies
North Eastern Regional Institute of Science and Technology, Nirjuli, Arunachal Pradesh

Rahul Dhiman
Gulzar Group of Institutes
Ludhiana, Punjab, India

Gourab Dutta
Department of Computational Sciences
Brainware University
Kolkata, West Bengal, India

Nabanita Ghosh
Christ University
Bengaluru, Karnataka, India

Rahul Kumar Ghosh
Department of Computational Sciences
Brainware University,
Kolkata, West Bengal, India

R. Gowri
Department of Computer Science and Engineering
KIT-Kalaignarkarunanidhi Institute of Technology
Kannampalayam, Tamil Nadu, India

Jyoti Hatiboruah
Department of Business Administration
University of Science and Technology, Meghalaya
Ri-Bhoi, Meghalaya, India

K. Jagadeesh
Department of AIDS
SNS College of Engineering
Coimbatore, Tamil Nadu, India

Vinay Jain
Department of Pharmacognosy
ShriRam College of Pharmacy
Banmore, Morena, Madhya Pradesh, India

Subrata Jana
Department of Mathematics
Jadavpur University
Kolkata, West Bengal, India

Jeswin Jose
Department of Commerce
Christ University
Bengaluru, Karnataka, India

K. Lohith
Consultant—US Practitioner
Infosys Limited
Houston, Texas, USA

Abhijit Mohanty
Electronics & Instrumentation Engineering
Silicon Institute of Technology
Silicon University
Bhubaneswar, Odisha, India

Anita Mohanty
Electronics Engineering
Silicon Institute of Technology
Silicon University
Bhubaneswar, Odisha, India

Subrat Kumar Mohanty
Electronics and Computer Engineering
Krupajal Engineering College
Biju Patnaik University of Technology
Bhubaneswar, Odisha, India

Ambarish G. Mohapatra
Electronics Engineering
Silicon Institute of Technology
Silicon University
Bhubaneswar, Odisha, India

Y. Mukhammadmurod
Director of Academic Planning and Development
Silk Road International University of Tourism and Cultural Heritage
Samarkand, Uzbekistan

Anshit Mukherjee
Department of Computer Science
Abacus Institute of Engineering and Management
Mogra, West Bengal, India

Gunjan Mukherjee
Department of Computational Science
Brainware University
North 24 Parganas, West Bengal, India

Simran Hemanshu Naik
Department of Commerce
Christ University
Bengaluru, Karnataka, India

K. Kritin Nanjappa
Department of Commerce
Christ University
Bengaluru, Karnataka, India

Rajive Mohan Pant
Vice Chancellor
Assam University
Silchar, Assam, India

S. Parul
Gulzar Group of Institutions
Ludhiana, Punjab, India

C. Ramesh
Department of Computer Science and Engineering
KIT-Kalaignarkarunanidhi Institute of Technology
Kannampalayam, Tamil Nadu, India

Sunil Kumar RM
Department of Computer Science and Engineering
R.L. Jalappa Institute of Technology
Bangalore, Karnataka, India

Soumen Santra
Department of Master of Computer Application
Techno International New Town
Kolkata, West Bengal, India

Anirban Sarkar
Department of Management and Marketing
West Bengal State University
North 24 Parganas, Berunanpukhuria, West Bengal, India

Mohammad Shahparan
Department of Tourism
Silk Road International University of Tourism and Cultural Heritage
Samarkand, Uzbekistan

Pankaj Sharma
Department of Pharmaceutics
ShriRam College of Pharmacy
Morena, Madhya Pradesh, India

B. Shuriya
Department of Computer Science and Engineering
United Institute of Technology
Coimbatore, Tamil Nadu, India

S. Simonthomas
Aarupadai Veedu Institute of Technology
Vinayaka Mission's Research Foundation (Deemed to be University)
Chennai, Tamil Nadu, India

K. Taranjeet
Gulzar Group of Institutions
Ludhiana, Punjab, India

Tran Minh Tung
FPT School of Business and Technology (FSB)
FPT University
Vietnam

M. Venus
Gulzar Group of Institutions
Ludhiana, Punjab, India

K. Yogeshwaran
Department of Computer Science and Engineering
KIT-Kalaignarkarunanidhi Institute of Technology
Kannampalayam, Tamil Nadu, India

1 Fundamentals of Fuzzy Logic
Theory and Applications

Anita Mohanty, Ambarish G. Mohapatra, Subrat Kumar Mohanty, and Abhijit Mohanty

1.1 INTRODUCTION

Fuzzy logic is a powerful mathematical framework that extends classical logic to take care of uncertainty and imprecision, making it a cornerstone of modern computational intelligence. Originating from the work of Lotfi Zadeh in the 1960s, fuzzy logic provides a systematic approach to reasoning with approximate or vague information, which is often encountered in real-world scenarios [1], as shown in Figure 1.1.

Unlike conventional two-valued logic handles truth values that are also true or false, fuzzy logic handles truth values that range from 0 to 1. These functions are especially significant for disciplines like control systems, knowledge of which can readily adjust to various conditions and inputs that typical quantitative methods cannot define. The fuzzy logic consists of basic components like fuzzy sets, membership functions, and fuzzy rules—fundamental tools in modeling and solving real systems. Fuzzy logic is implemented in a wide range of applications across automotive systems, health care, finance, and robotics among others, thus making it very useful in today's world. Unlike traditional techniques, binary logic recognizes only true and false, fuzzy logic is more practical in tackling real-life difficulties insofar as the difficulties are marked by uncertainty and variability [2].

1.1.1 Overview of Fuzzy Logic

Fuzzy logic is an extension of classical binary forms of language with the addition of a truth scale, to address the vagueness of the world. Unlike mathematical digital sets or binaries where the variables are classified either as 1 or as 0, fuzzy logic allows variables to be interpolated between 0 and 1. This somewhat flexible approach makes fuzzy logic even more appropriate for capturing the behaviors of systems that cannot be described adequately using just dichotomous logic modes [3].

DOI: 10.1201/9781003538837-1

FIGURE 1.1 Concept of fuzzy logic.

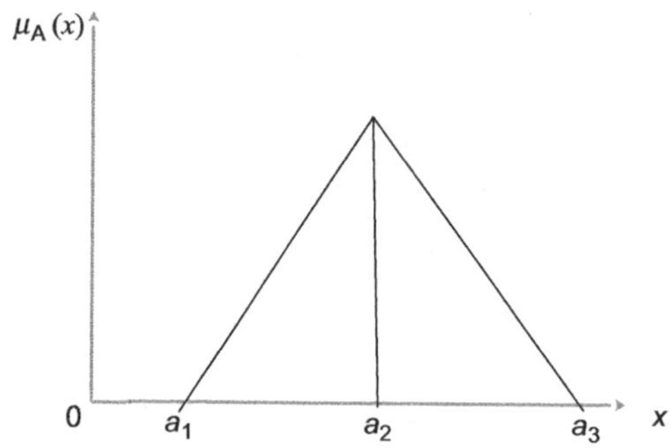

FIGURE 1.2 Triangular membership function.

1. Fuzzy Sets and Membership Functions

In fuzzy logic, the most important term is in the form of *fuzzy sets*, for each element was given an attributive or membership, ranging from 0 to 1. This membership is computed using *membership functions* that give the extent of membership of an input variable to a fuzzy set.

Example Membership Function: The commonly used triangular membership function is defined as follows:

> A *triangular membership function* is well-defined by three parameters a_1, a_2, and a_3 that define the position of the triangle concerning the base and the apex of the triangle, as illustrated in Figure 1.2.

The function $\mu_A(x)$ describes the degree of membership of x in the fuzzy set A in Eq. (1.1):

$$\mu_A(x) = \begin{cases} 0 & \text{if } x \le a_1 \\ \dfrac{x - a_1}{a_2 - a_1} & \text{if } a_1 < x \le a_2 \\ \dfrac{a_3 - x}{a_3 - a_2} & \text{if } a_2 < x \le a_3 \\ 0 & \text{if } x > a_3 \end{cases} \tag{1.1}$$

a_1: Left endpoint of the triangle where the membership is 0.
a_2: A peak of the triangle where the membership is 1.
a_3: Right endpoint of the triangle where the membership is again 0.

2. Fuzzy Rules and Inference

In fuzzy logic, rules like "IF temperature is High THEN fan speed is Fast" are used to infer actions based on the degree of truth. These rules help translate fuzzy concepts into actionable outcomes.

3. Fuzzy Operations

Fuzzy operations generalize traditional logic operations to handle degrees of membership. For example: **AND Operation,** which is the minimum operator that is used to find the intersection of fuzzy sets, as shown in Eq. (1.2).

$$\text{A AND B} = \min\left(\mu_A(x), \mu_A(x)\right) \tag{1.2}$$

Here $\mu_A(x)$ and $\mu_B(x)$ are the membership functions for fuzzy sets A and B, respectively.

4. Example Table of Membership Values

Table 1.1 with membership values demonstrates how different temperatures are mapped to their degree of membership in a fuzzy set "High."

TABLE 1.1
Mapping of Different Temperature Values to Degree of Membership

Temperature (°C)	Membership in "High"
15	0
20	0.4
25	1
30	0.8
35	0

5. Defuzzification

The fuzzy values result in a crisp value by using defuzzification. The *centroid method* represented by Eq. (1.3) is a common technique:

$$\text{Output} = \frac{\int x \cdot \mu_{\text{output}}(x)\,dx}{\int \mu_{\text{output}}(x)\,dx} \tag{1.3}$$

$\mu_{\text{output}}(x)$: The membership function of the output fuzzy set
$\int x \cdot \mu_{\text{output}}(x)\,dx$: The weighted sum of output values
$\int \mu_{\text{output}}(x)\,dx$: The total membership weight

These equations and methods provide a formal basis for working with fuzzy logic in various applications.

1.1.2 Historical Background and Development

Fuzzy logic is an approach to handling vagueness and ambiguity in information and decision-making and was developed by Lotfi Zadeh in 1965. Its development is profoundly a breakaway from conventional classical binary logic, which can only have variables that take the value true or false. However, fuzzy logic can only consider true or false values between 0 and 1 since the world operates with little gray between black and white.

1.1.2.1 Early Foundations

The genesis for fuzzy logic is attributed to Lotfi Zadeh in his paper "Fuzzy Sets," published in 1965. The idea of fuzzy sets as a way to represent and manipulate vague or imprecise information was introduced by Zadeh. Unlike classical sets, which are binary (an element either belongs or does not belong to a set), fuzzy sets allow partial membership. This concept was a radical shift, laying the groundwork for a new approach to logic and decision-making [4].

Development through the Decades (Figure 1.3)

- **1960s–1970s**: The early research focused on the theoretical aspects of fuzzy logic and fuzzy sets. Zadeh's work on fuzzy sets and fuzzy logic was pioneering, but its practical applications were initially limited. The concept was primarily explored in academic circles and theoretical studies.
- **1980s**: The 1980s saw the emergence of fuzzy logic in practical applications. The growth of fuzzy control systems, such as those used in washing machines and air conditioners, demonstrated the real-world utility of fuzzy logic. This period marked the beginning of fuzzy logic's transition from theory to practice.
- **1990s**: The 1990s were characterized by the expansion of fuzzy logic into various industries, including automotive, aerospace, and industrial control. Fuzzy logic controllers (FLCs) and fuzzy expert systems have become

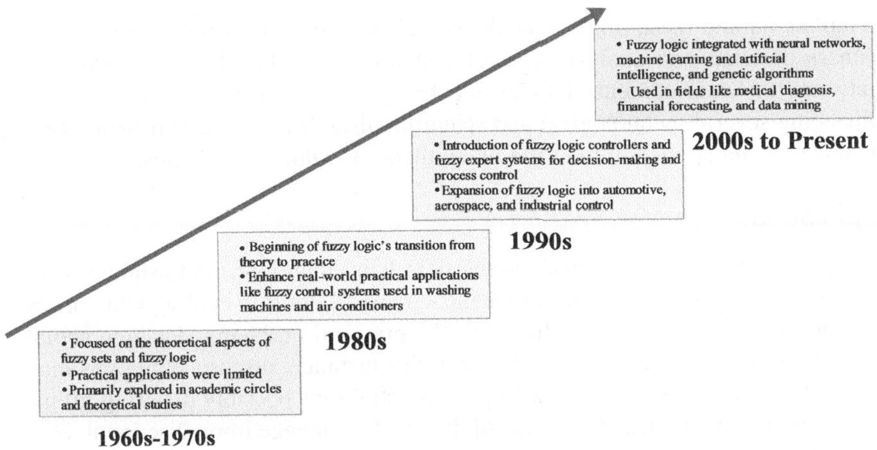

FIGURE 1.3 Development of fuzzy logic through decades.

popular in decision-making and process control due to the new development. They were claimed to be automotive systems such as automatic transmission, home appliances such as washing machines, and industrial applications.

- **2000s to Present**: The 21st century has seen many improvements in the use of fuzzy logic with other computational intelligence tools like neural networks as well as genetic algorithms. Fuzzy logic is used in irregular areas like medical diagnosis, financial forecasting, data mining, and many more. Recent developments are presented with the integration of fuzzy logic with machine learning and artificial intelligence.

Key Milestones and Contributions

- **1965**: Lotfi Zadeh publishes "Fuzzy Sets," introducing the concept of fuzzy logic.
- **1970s**: Initial applications of fuzzy logic in industrial and control systems begin to emerge.
- **1980s**: Development and commercialization of fuzzy logic controllers in consumer electronics and automobile companies.
- **1990s**: Expansion into diverse fields, including expert systems and medical diagnosis.
- **2000s to Present**: Integration with AI techniques and expansion into data analytics, finance, and health care.

1.1.3 Importance and Relevance in Modern Applications

Fuzzy logic has become a basic element in the tools applied in systems utilizing uncertainty and imprecision. The feature that has seen it used widely is its capacity to emulate human reasoning and decision-making, hence useful in control systems, business decision and analysis support systems, and data interchange [5]. Uncertainty and the development of fuzzy set theory and fuzzy logic further propagate influencing several technological and scientific disciplines, which enshrine the sustainableness and applicability of the fuzzy theory to address problems.

Applications

- *Handling uncertainty and complexity:* It is, therefore, understandable to consider fuzzy logic as one of the most powerful tools in dealing with uncertainty in information. In almost all the practical problems, data and actual life experiences may not be clear-cut. For instance, disease symptoms may not be unique and distinct with patient conditions possibly differing infinitesimally in medical diagnosis [6]. It can also manage imprecise input data, which conventional systems are not capable of doing, as they only work in blacks and whites, while fuzzy logic systems can provide more decisions that are precise.
- *Applications in control systems:* In many fields of control systems, fuzzy logic is used. In industrial automation, fuzzy logic controllers are used for controlling complex dynamic processes that need self-tuning and self-adjustability [7]. For instance, in automotive systems, fuzzy logic is applied in choosing the best mode of cylinder ignition, proper management of transmission, and proper handling of the active suspensions. In the same way, fuzzy logic within consumer electronics is utilized within washing machines to adapt rinse settings according to the load capacity and even the level of dirt, to improve the device's performance and utility for the consumer.
- *Expert systems and decision support:* Fuzzy logic as a technology tool has been considered an important tool for supporting the creation of expert systems and decision support systems. These systems employ fuzzy logic to mimic human resources in decision-making activities, which are difficult to quantify [8]. For instance, in business forecasting fuzzy logic can be employed in trend analysis recommended investment opportunities under conditions of uncertainty. In other cases, such as in the health care field, fuzzy logic helps in medical diagnosis and planning, the subsequent action to be taken because of symptoms or history or some diagnostic information received.
- *Data classification and pattern recognition:* In the so-called data mining, fuzzy logic becomes particularly important in the classification and pattern identification. It is especially beneficial in cases when the categories of data are rather overlapping or not particularly distinguishable. For instance, in image processing, fuzzy logic is helpful in the proper classification of pictures depending on ill-defined characteristics like texture or color, which can enhance the value of a pattern recognition system [9]. In the same way, in NLP, fuzzy logic helps in the analysis and processing of the necessary language, which is ill-defined and depends on context.

- *Case studies and real-world implementations:* Indeed, there are many examples in practice of how the power and efficiency of fuzzy logic is evident. For example, in agriculture, fuzzy-owned methods are applied in smart irrigation systems to decide the amount of water needed and when it is needed most by checking the moisture status of the soil and the prevailing weather conditions [10]. Fuzzy logic applied in environmental monitoring helps in making a more accurate analysis of the degree of pollution and impacts. In robotics, fuzzy logic allows the robots to operate and control in conditions where the environment is unpredictable.
- *Adaptability and future prospects:* Due to this, fuzzy logic could be applicable in several subbranches of almost all emerging disciplines. Many artificial intelligence applications are incorporating fuzzy logic, and computer learning and data analytics are advancing with the technology boost. Further development may encompass issues on the enhancement of hybrid fuzzy systems where fuzzy logic is incorporated with even other computational models to analyze more complicated and changing conditions [11].

Therefore, as the result remains open-ended and the complexity of the systems is manageable, fuzzy logic remains a very useful instrument in contemporary applications. It applies to different fields, including industrial applications, and data analysis showing evidence of future developmental possibilities.

1.2 FUNDAMENTAL CONCEPTS

The componential parts of fuzzy logic are fuzzy sets, membership functions, and linguistic variables, which enshrine the measures for the handling of uncertain and imprecise data. The definition of fuzzy sets is an advancement of classical binary logic by the incorporation of the membership degree that is characterized by membership functions, including triangular membership and Gaussian functions. These functions expressed the degree of an element in the set by a fuzzy set, thus facilitating a richer representation of data compared to the crisp set [12]. Linguistic variables define these fuzzy sets in terms of natural language, thereby enabling imitation of the human expert system in rule-based reasoning. With these components, fuzzy logic systems can provide the best and most flexible way of implementing complicated nonlinear real-world phenomena than that offered by a system of classical crisp logic.

1.2.1 Fuzzy Sets and Membership Functions

Fuzzy sets are among the key concepts in fuzzy logic, which generalizes the notion of the set in mathematics to consider the problem of vagueness. In contrast to classical sets, where an element either belongs or does not belong to the set, fuzzy sets allow for partial membership, where an element can partially belong to the set to varying degrees [13].

Membership functions describe how each element in the input space is projected to a membership value between 1 and 0. This function provides a calculation of the

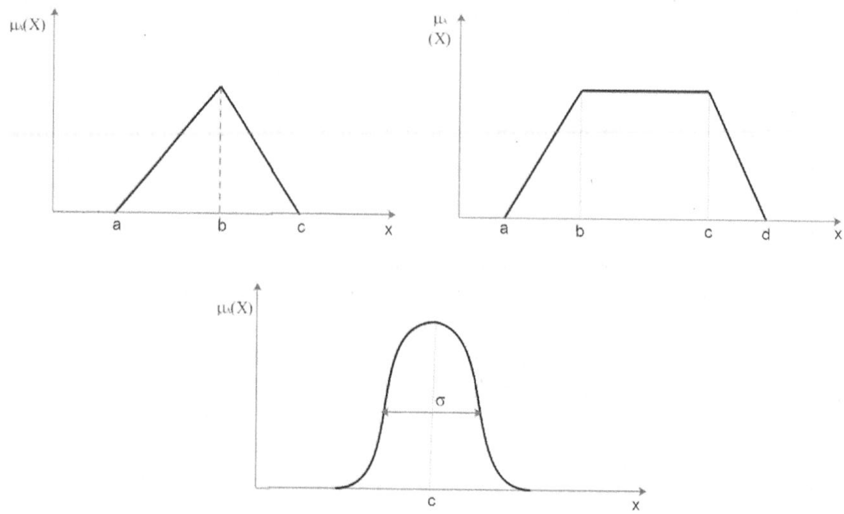

FIGURE 1.4 Different types of membership functions.

degree of membership of each element in the fuzzy set. Mostly used membership functions are shown in Figure 1.4.

- *Triangular membership function:* This very widespread membership function has a triangular form. It is characterized by three constants *a*, *b*, and *c*, where *a*, *b*, and *c* are the "feet" of the triangular form and *b* is the "peak." The membership function $\mu_A(x)$ for a triangular fuzzy set A depends on Eq. (1.4):

$$\mu_A(x) = \begin{cases} \frac{x-a}{b-a} & \text{if } a \leq x \leq b \\ \frac{c-x}{c-b} & \text{if } b \leq x \leq c \\ 0 & \text{otherwise} \end{cases} \tag{1.4}$$

- *Gaussian membership function:* Another such function is the Gaussian membership function, characterized by its smooth bell-shaped curve. It is defined by a mean *c* and a standard deviation σ, and is given by Eq. (1.5):

$$\mu_A(x) = \exp\left(-\frac{(x-c)^2}{2\sigma^2}\right) \tag{1.5}$$

This function is applicable where fuzzy sets are required, with degrees of membership ranging from 0 to 1; it is appropriate for smooth distribution curves.

- *Trapezoidal membership function:* Trapezoidal membership function is very much related to the triangular function but has a flat top. It is defined by four parameters *a*, *b*, *c*, and *d*, where *a* and *d* represent the "feet," and *b* and *c* represent the "shoulders" of the trapezoid. The trapezoidal membership function $\mu_A(x)$ for a set A is given by Eq. (1.6):

$$\mu_A(x) = \begin{cases} \dfrac{x-a}{b-a} & \text{if } a \leq x \leq b \\ 1 & \text{if } b \leq x \leq c \\ \dfrac{d-x}{d-c} & \text{if } c \leq x \leq d \\ 0 & \text{otherwise} \end{cases} \tag{1.6}$$

This function is used for fuzzification of the sets in cases where a range of values fully belong to that set; it thus has a wider transition than the triangular membership function.

1.2.2 Comparison with Classical Binary Logic

Classical binary logic deals with binary values, 0 and 1, and has its underlying concept based on the crisp set theory. An element is either completely a member of a set or it is not a member of that set at all, membership values are either 1 or 0 [3].

Fuzzy logic, on the other hand, can incorporate a partial membership depicted in a range between 0 and 1. It increases the reality in which organizational relationships are often not clearly defined and are rather fluid.

- **Classical Set Example**
  - If set A is defined as the set of "tall people," then a person is either in set A (tall) or not (not tall).
- **Fuzzy Set Example**
  - In a fuzzy set of "tall people," a person might have a membership value of 0.7, indicating partial membership; this means the person is somewhat tall.

An example of a comparison of classical and fuzzy memberships is shown in Table 1.2.

1.2.3 Linguistic Variables and Their Use in Fuzzy Logic

Linguistic variables are those variables for which a value may be any word or phrase in natural or artificial language, rather than a numeral. They offer means by which the qualitative data can be encoded and processed using in fuzzy logic system [14].

TABLE 1.2
Comparison of Classical and Fuzzy Memberships

Attribute	Classical Logic	Fuzzy Logic
Membership value	0 or 1	0 to 1
Example	Tall (1)/not tall (0)	Tall (0.7)
Precision	Crisp, binary	Gradual, nuanced

TABLE 1.3
Linguistic Terms Expressing Fuzzy Set Parameters

Linguistic Term	Fuzzy Set Parameters (e.g., for Temperature)
Low	$\text{Temp} \leq 20$
Medium	$20 < \text{Temp} \leq 30$
High	$\text{Temp} > 30$

- *Linguistic terms:* These are terms applied in fuzzy logic to mean fuzzy sets of a particular object. For instance, the fuzzy set can be high, medium, and low to qualify the temperatures.
- *Fuzzy rules:* Linguistic variables are employed in the formulation of fuzzy rules, the pattern of the statements used being the IF-THEN type. For example:
  - *Rule 1:* IF temperature is High THEN fan speed is Fast.
  - *Rule 2:* IF temperature is Medium THEN fan speed is Medium.

Table 1.3 describes linguistic variables in the form of an example.

1.2.3.1 Use in Fuzzy Systems

Linguistic variables enable the system to operate with qualitative data and provide its answers in the form of decision-making based on human heuristics. These systems enable data to be translated into fuzzy sets and rules and therefore ones that analyze data in a way that is very similar to the human brain.

1.3 MATHEMATICAL FOUNDATIONS

The algebraic structures of fuzzy logic systems represent the basic computations and features that govern the working of fuzzy sets and relations. Some of those include union, intersection, and complement operations, which generalize classical set operations with those having degrees of memberships. Fuzzy relations are the measures of relatedness between two elements, one from each sets, while the logical connectors apply relations to produce compound relations. Also, fuzzy arithmetic

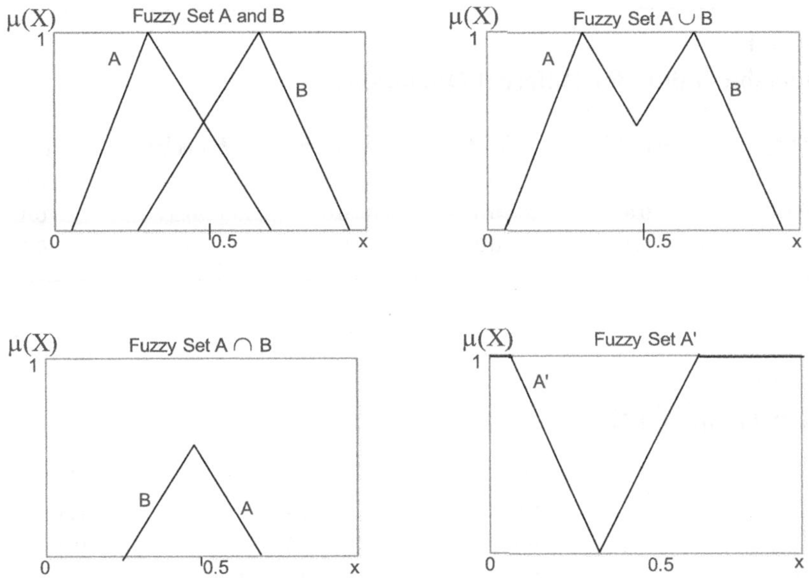

FIGURE 1.5 Different operations on fuzzy sets.

helps in making calculations regarding fuzzy numbers which are defined with membership functions such as triangular or trapezoidal forms [15]. They can help in how to model and interpret uncertainty and imprecision in a number of applications and offer strong stepping stones for using fuzzy logic in practice.

1.3.1 Operations on Fuzzy Sets: Union, Intersection, Complement

Fuzzy set operations follow up the classical set operations but are able to deal with the case when an element belongs to a set to a certain degree, which can be any real number between 0 and 1. The primary operations are shown in Figure 1.5.

- **Union**: The membership function of the union of two fuzzy sets A and B is given in Eq. (1.7):

$$\mu_{A \cup B}(x) = \max\left(\mu_A(x), \mu_B(x)\right) \tag{1.7}$$

  $\mu_A(x)$ and $\mu_B(x)$ represent the membership functions of fuzzy sets A and B, respectively.
- **Intersection**: The membership function of the intersection of two fuzzy sets A and B is given in Eq. (1.8):

$$\mu_{A \cap B}(x) = \min\left(\mu_A(x), \mu_B(x)\right) \tag{1.8}$$

TABLE 1.4
Membership Values for Different Operations

Element x	$\mu_A(x)$	$\mu_B(x)$	$\mu_{A\cup B}(x)$	$\mu_{A\cap B}(x)$	$\mu_{A^c}(x)$
1	0.4	0.7	0.7	0.4	0.6
2	0.6	0.5	0.6	0.5	0.4
3	0.9	0.2	0.9	0.2	0.1

TABLE 1.5
Fuzzy Relation Matrix

R	b_1	b_2	b_3
a_1	0.6	0.8	0.5
a_2	0.7	0.5	0.9
a_3	0.3	0.6	0.4

- **Complement**: The membership function of the complement of a fuzzy set A is given in Eq. (1.9):

$$\mu_{A^c}(x) = 1 - \mu_A(x) \tag{1.9}$$

Examples of different operations that are possible for different membership functions are given in Table 1.4.

1.3.2 Fuzzy Relations and Compositions

Fuzzy Relations: A fuzzy relation R between two fuzzy sets A and B is described by a fuzzy relation matrix where each element r_{ij} denotes the degree of relationship between elements a_i and b_j, as given in Table 1.5. The degree of relationship between a_1 and b_1 is 0.6, and between a_2 and b_3 is 0.9.

The membership function of the fuzzy relation R is given in Eq. (1.10):

$$\mu_R(a,b) = \min\left(\mu_a(x), \mu_b(x)\right) \tag{1.10}$$

Composition: Fuzzy composition of relations combines two fuzzy relations R_1 and R_2 into a new relation R. If R_1 is a relation from set X to set Y and R_2 is a relation from set Y to set Z, the composition R from X to Z is defined by Eq. (1.11):

$$\mu_R(x,z) = \sup_{y\in Y}\min\left(\mu_{R_1}(x,y), \mu_{R_2}(y,z)\right) \tag{1.11}$$

1.3.3 Fuzzy Arithmetic and Fuzzy Numbers

Fuzzy Arithmetic: Fuzzy arithmetic operations extend classical arithmetic to fuzzy numbers. For example, the sum of two fuzzy numbers A and B, represented by their membership functions $\mu_A(x)$ and $\mu_B(x)$, is defined by Eq. (1.12):

$$\mu_{A+B}(x) = \sup_{a+b=x} \min(\mu_A(a), \mu_B(b)) \tag{1.12}$$

Fuzzy Numbers: Fuzzy numbers are represented by membership functions, which show levels of partial truth of validity. There are two basic forms of fuzzy numbers: triangular and trapezoidal.

1.4 FUZZY INFERENCE SYSTEMS

Fuzzy inference systems (FIS) are the main computational structure of fuzzy systems acting as an engine for making decision within the context of a given set of rules and principles. An FIS works through inputs that translate into outputs with "If-Then" rule base that define the semantics of the fuzzy sets of variables. Fuzzy inputs are managed using fuzzification; fuzzy rules are applied by inference and a fuzzy output is generated [16, 17]. Lastly, defuzzification is carried out to obtain the crisp value to be implemented after the FIFO had been made fuzzy. FIS is applied in control systems, pattern recognition, decision support systems, etc.; it realizes the management of uncertainty and vagueness.

1.4.1 Structure and Components of Fuzzy Inference Systems

A fuzzy inference system structure consists of several components, as shown in Figure 1.6.

- *Fuzzification:* Transforms sharp input values into fuzzy sets. This is accomplished with the membership functions, which transform the input values into belongingness levels, with a value between 0 and 1.
- *Knowledge base:* Includes a set of fuzzy rules, which have been defined in terms of the functions of membership. They are expected in the form of IF-THEN format most of the time.
- *Inference engine:* Blurs up the input data to get a fuzzy output based upon the rankings preserved in the knowledge base and applied by the fuzzy rules.
- *Defuzzification:* Transforms the fuzzy result back into a clear number, normally by centroid or weighted average methods.

1.4.2 Fuzzy Rule-Based Systems

A fuzzy rule-based system (FRBS) is an artificial intelligence model developed to mimic specific decision-making organizations or systems where the decision-making

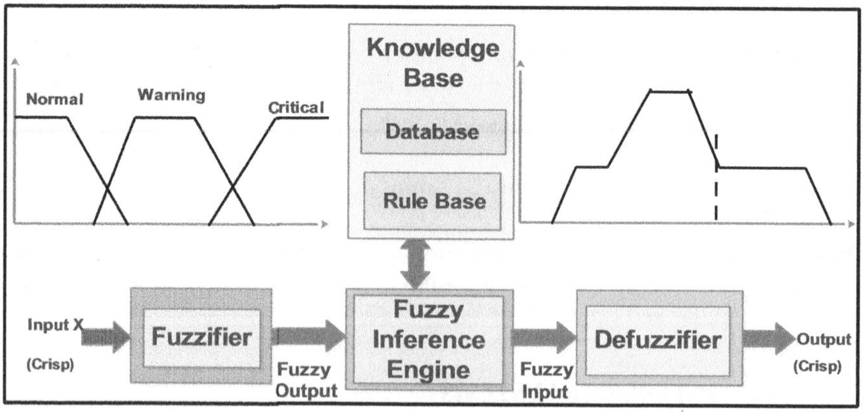

FIGURE 1.6 Structure of a fuzzy inference system.

processes are difficult to articulate. These rules are expressed in the form of "IF-THEN" statements, for example:

Rule 1: IF temperature is High THEN fan speed is Fast
Rule 2: IF temperature is Medium THEN fan speed is Medium

The Min–Max composition is among the techniques used in the assessment of these rules. Given fuzzy sets A and B, the membership function for the rule "IF A THEN B" can be calculated using Eq. (1.13):

$$\mu_R(x,y) = \min\left(\mu_A(x), \mu_B(y)\right) \tag{1.13}$$

where $\mu_A(x)$ is the membership value of x in fuzzy set A.
$\mu_B(y)$ is the membership value of y in fuzzy set B.

1.4.3 Design and Implementation of Fuzzy Inference Systems

FIS design entails the identification of antecedent and consequent input and output variables, the definition of MFs, and the development of fuzzy rules. For example, let there be a FIS to be used in regulating a fan with an input of temperature. The input (temperature) can be divided into three fuzzy sets: low, medium, and high. The output (fan speed) can also have three fuzzy sets: they indicated slow, medium, and fast, as presented in Table 1.6.

These membership values are used in the FIS to determine the output fan speed based on the input temperature. The defuzzification process converts the fuzzy output into a crisp value, using the centroid method given in Eq. (1.14):

$$\text{Crisp output} = \frac{\sum_i \mu_i . x_i}{\sum_i \mu_i} \tag{1.14}$$

TABLE 1.6
Membership Values for Temperature and Fan Speed

Temperature (°C)	Low	Medium	High
15	1.0	0.0	0.0
20	0.4	0.6	0.0
25	0.0	1.0	0.0
30	0.0	0.6	0.4
35	0.0	0.0	1.0

where μ_i is the membership value of the output fuzzy set.
x_i is the corresponding value of the output variable.

This leads to an accurate control output, which may be employed to exercise the fan within a real environment.

1.5 APPLICATIONS OF FUZZY LOGIC

Fuzzy logic has versatility in a wide range of fields because it contains flexibility in dealing with uncertainty and is based on an approximate reasoning system. Fuzzy logic is applied in control systems resulting in formation of fuzzy controller that can handle hardened processes, for instance, in systems like air-conditioning systems and in automation of several industrial operations. Fuzzy logic is especially important in decision-making in absence of information; it is used in medical diagnosis, financial analysis, and risk evaluation [18]. Furthermore, in data classification and pattern recognition, fuzzy logic is also used in that data, which can be classified, cannot be easily defined; fuzzy logic is then used to cluster it. Examples of fuzzy sets in the real world show their applications from simple general use, for example, washing machines and cameras to sophisticated complex systems such as robotics and artificial intelligence, where it improves the system's robustness and effectiveness in real-life applications.

1.5.1 Control Systems (e.g., Fuzzy Controllers in Automation)

It should also be noted that in the control systems, fuzzy logic is used widely for the management of complex and nonlinear processes. An example application area is fuzzy controllers for automation where regular control techniques can perform inadequately. These controllers are used to make decisions where the inputs are inexact; these give a bit-by-bit control of the system [19]. For example, in an HVAC system, a fuzzy controller may alter the temperature by using variables such as cool, warm, or hot, instead of new exact temperature measurements. The fuzzy control rules are typically structured in "IF-THEN" statements, for example,

IF the temperature is "cool" THEN the heating power is "low".

It enables a more human-like reasoning and thus increases the effectiveness of the performance of the system.

1.5.2 Expert Systems (e.g., Medical Diagnosis, Financial Decision-Making)

About decision-making, fuzzy logic is the primary means by which uncertain or difficult to model data can be utilized in an expert system. For instance, in a disease diagnosis, fuzzy logic diagnoses the diseases with symptoms and patient data, which are generally ambiguous or seek basic information [20]. The rules might look like as follows:

> IF symptom A is "mild" AND symptom B is "severe" THEN the probability of disease X is "high".

This enables the system to give suggestions in case the obtained input data is ambiguous. Likewise, in financial decision-making, fuzzy logic will be useful to provide the numerical model of markets and risks required for investment methods for conditions that cannot be defined clearly.

1.5.3 Data Classification and Pattern Recognition

Fuzzy logic also proves to be effective in data classification and pattern recognition especially when the data points do not fit certain classes. An example is fuzzy clustering, which enables a datum to belong to more than one cluster, with specific membership level. The membership function $\mu(x)$ gives the belongingness of an input data x to the particular cluster [21]. This proves useful especially in image and voice analysis because data can sometimes be confusing and contaminated by noise. Table 1.7 represents the membership values for different data points in a fuzzy clustering algorithm.

1.5.4 Case Studies: Real-World Implementations of Fuzzy Logic

Various applications of fuzzy logic have been practiced on actual systems in several fields of specialization. For example, fuzzy logic controllers are applied in the automotive industry, for instance, to control automatic transmission with an objective of identifying proper gear shifting retrospect to some input variables like load and speed [22]. The examples from the consumer electronics domain include the washing machines that employ fuzzy logic to modify the washing cycle corresponding to size of the load and fabric type. Another example of its application is robotics.

TABLE 1.7
Membership Values for Different Data Points

Data Point	Cluster 1 Membership	Cluster 2 Membership
Point A	0.7	0.3
Point B	0.4	0.6
Point C	0.9	0.1

In uncertain environment, fuzzy logic assists robots to move and to make decisions. By use of case studies in this chapter, there is a demonstration of how fuzzy logic is useful in improving the flexibility, stability, and efficiency of a system it is applied to, thus making it vial in the current developed technology.

1.6 ADVANTAGES AND LIMITATIONS

The rationale behind fuzzy logic is to produce an improved method for processing unprecise and vague information as compared to the classical binary system where decisions are often poor or poorly defined; it is widely applied to decision-making and control systems. It has application in automation, medical diagnosis, and pattern recognition where ambiguous data is characteristic of the true model. However, generalizability using IMF also has its drawbacks, as previously discussed [23]. Fuzzy systems are not very rigorous or "cold" because the design process of fuzzy systems can be charged with a certain level of subjectivity. Further, with increased and different inputs and multiple rules, the entire system becomes cumbersome and more confusing when it comes to processing of results. However, due to the flexibility of fuzzy logic, and its effective and efficient application to various fields, it continues to spread, mostly when used in combination with other computational methods, in order to overcome the above-stated drawbacks.

1.6.1 Strengths of Fuzzy Logic in Handling Uncertainty and Vagueness

Fuzzy logic is least precise but most suitable where certainty and accuracy measurements are easy to define but difficult to quantize. Fuzzy logic is more elastic as compared to the two-valued logic as it incorporates true and false on a continuum rather than Boolean algebra propositions or propositions with values between 0 and 1. For instance, while defining tall people, the domain fuzzy set of tall people will be in the range of 0 and 1 and will contain somewhat tall, very tall, and moderately tall individuals. This flexibility makes fuzzy logic very efficient in applications such as climate control, decision support systems, and pattern recognition because the input data is generally very imprecise.

Table 1.8 shows how different heights map to the membership value in the fuzzy set "Tall." For instance, a height of 170 cm might be considered moderately tall with a membership value of 0.5.

TABLE 1.8
Mapping of Height into Membership Values

Height (cm)	Membership in "Tall"
150	0.0
160	0.2
170	0.5
180	0.8
190	1.0

TABLE 1.9
Key Difference between Fuzzy Logic and Probabilistic Methods

Aspect	Fuzzy Logic	Probabilistic Methods
Basis	Degree of membership in fuzzy sets	Likelihood of events based on probability distribution
Data requirement	Does not require a large dataset; subjective	Requires a well-defined probability distribution
Uncertainty handling	Handles uncertainty through degrees of membership	Handles uncertainty through statistical probabilities
Suitability	Qualitative reasoning, human-like decision-making	Quantitative reasoning, risk assessment
Complexity	Can be subjective and complex with many inputs and rules	Based on mathematical models, complexity depends on the data
Application scenarios	Control systems, pattern recognition, decision-making	Risk assessment, financial modeling, predictive analytics

1.6.2 Comparison with Other Approaches (e.g., Probabilistic Methods)

Unlike probabilistic methods, fuzzy logic does not require a comprehensive statistical uncertainty model. While probabilistic methods rely on the likelihood of events and require a well-defined probability distribution, fuzzy logic is based on the degree of membership in a fuzzy set, which can be subjective and does not necessarily require a large dataset [24]. However, probabilistic approaches excel in scenarios where precise probabilities can be derived, such as in risk assessment. Fuzzy logic, on the other hand, is more suitable for qualitative reasoning and situations where human-like reasoning is needed.

Table 1.9 highlights the key differences between fuzzy logic and probabilistic methods, emphasizing their respective strengths and applications.

1.6.3 Challenges/Limitations in Fuzzy Logic Applications

Table 1.10 outlines key issues that practitioners may encounter when applying fuzzy logic, emphasizing the need for careful consideration in system design and implementation.

1.7 FUTURE TRENDS AND RESEARCH DIRECTIONS

New areas of fuzzy logic research appear to have promising further development since fuzzy logic remains an active area of research and expansion of its applications in combination with other computational intelligence methods. Some new exciting fields are in the intersection with deep learning like autonomous vehicle systems, smart energy systems, and smart robot systems for which heuristic methods to learn from experience are quite beneficial. Recent developments in research aim to integrate fuzzy logic systems by incorporating neural and evaporative algorithms and

TABLE 1.10
Challenges/Limitations in Fuzzy Logic Applications

Challenges/Limitations	Description
Design complexity	Designing fuzzy logic systems requires careful selection of membership functions and rules, which can be complex and subjective.
Management of complexity	The complexity of fuzzy systems increases with the number of inputs and rules, making management and maintenance more difficult.
Lack of standardization	The general outcome of defuzzification is dependent on the type of technique used, and as yet there is no agreed-upon standard on the type or strategy of defuzzification to be used.
Time-consuming development	The construction of the fuzzy logic systems and the refinement process can be labor-intensive, principally in circumstances where many rules and memberships are to be implemented.
Scalability issues	If fuzzy systems are extended to larger or more complicated applications, the degree of computational complexity and possible performance degradation may arise.
Subjectivity in design	The problem with the modeling of the spaces of the membership functions and fuzzy rules and determinations is that these procedures may be rather ambiguous and lead to inconsistency and unreliability of the system.

machine learning into the system to increase flexibility and learning requirements [25]. In addition, the theory of fuzzy logic is gradually developing small and complex constructions for processing dynamic or big data, which also have the potential to bring more effective solutions. These trends imply that the future in the application of fuzzy logic is bright and that these forms will continue to cope with an increasing number of new complex technological problems.

1.7.1 Emerging Areas of Application for Fuzzy Logic

Traditional fuzzy logic is used for new areas of AI development like automobiles, control of smart electricity grid, and the Internet of Things (IoT). Table 1.11 gives some applications, where fuzzy logic facilitates decision-making and system management in environments characterized by uncertainty and variability.

1.7.2 Integration with Other Computational Intelligence Techniques

The future of fuzzy logic lies in its integration with other computational intelligence techniques, such as neural networks, genetic algorithms, and deep learning. This integration, often referred to as hybrid systems, leverages the strengths of each technique to improve overall system performance. For instance, a neuro-fuzzy system

TABLE 1.11
Emerging Areas of Applications

Application Area	Role of Fuzzy Logic	Details
Autonomous vehicles	Managing dynamic environments and decision-making under uncertainty	Interprets sensor data, handles varying driving conditions, and makes real-time driving decisions
Smart grids	Integrating renewable energy sources and managing power generation and demand variability	Handles uncertainty in power generation, optimizes energy distribution, and balances supply and demand
Internet of Things (IoT)	Enabling intelligent systems that adapt to varying data and conditions	Manages diverse and fluctuating data inputs from various IoT devices, and improves system adaptability and responsiveness

TABLE 1.12
Comparison of Computational Intelligence Techniques

Technique	Strengths	Weaknesses	Typical Applications
Fuzzy logic	Handles uncertainty and vagueness well	Subjective rule and membership function design	Control systems, decision-making
Neural networks	Excellent at learning patterns from data	Requires large datasets, complex training	Image recognition, speech processing
Genetic algorithms	Effective for optimization problems	Computationally intensive, slow convergence	Scheduling, optimization tasks
Deep learning	High performance with large datasets	Requires substantial computational resources	Natural language processing, AI models

combines the learning capabilities of neural networks with the reasoning capabilities of fuzzy logic, enabling more robust and adaptive control systems. Such hybrid systems are particularly useful in complex applications like robotics, where multiple types of uncertainty need to be managed simultaneously. As these technologies evolve and integrate further, they promise to push the boundaries of what is achievable in intelligent systems, making them more adaptive, accurate, and capable of addressing a broader range of real-world challenges. Table 1.12 gives a clear comparison of different computational intelligence techniques in terms of strengths, weaknesses, and applications.

TABLE 1.13
Comparison of Fuzzy Logic Types

Fuzzy Logic Type	Description	Applications	Advantages
Fuzzy logic (type-1)	Typical fuzzy logic with crisp membership functions	Pattern recognition, control systems	Easier to implement, well-established
Fuzzy logic (type-2)	Fuzzy sets with fuzzy membership functions	Medical diagnosis, speech recognition	Better to handle high uncertainty, more nuanced modeling

1.7.3 Advances in Fuzzy Logic Theory and Practice

More development occurs due to its continuous improvements of fuzzy logic theory and systems; the modifications concentrate on the interpretability, scalability, and efficiency of the system. Newer techniques are type-2 fuzzy logic where it is possible to model uncertainty in uncertainty since, in this case, the fuzzy sets themselves are fuzzy. This approach is especially beneficial in tasks with a large variability, for example, voice control and disease identification. Further studies present the investigation of new and improved techniques for the fuzzy inference and defuzzification processes by which logic can easily be implemented in real-time operations. Table 1.13 summarizes the difference between various fuzzy logic types.

1.8 CONCLUSION

However, objective difficulties are still present as the design of a large and complex system and several defuzzification methods are not universal. More development on how the fuzzy logic coupled with other methods and the exploration of advanced theoretical mathematical models for the fuzzy system are expected to build up the improved characteristics of the fuzzy system in the future. Such strengths continue to present difficulties such as high complication in system design and the absence of standard approaches to defuzzification. This is an area of much research interest in the future, particularly in combining fuzzy logic with other computational methods and improving the theoretical understanding of fuzzy logic. As more industries adopt the new complicated technologies, the need to integrate fuzzy logic in consequently intelligent systems will continue to be more innovative and experimental.

1.8.1 Recap of Key Points

Fuzzy logic is effective in the identification of system behavior, as well as the solution of problems in situations that are characterized by set, uncertain, and imprecise conditions. It is more elastic, hence it can provide measured or quantifiable results instead of black-and-white outcomes. With the great possibility of approximations included

TABLE 1.14
Key Points of Fuzzy Logic

Aspect	Description
Definition	A formalism to manage information and knowledge about uncertainty and approximation using truth values
Flexibility	Enables reaching decisions with degrees of truth instead of Yes/No responses
Applications	Most popular in the area of control systems, expert systems, pattern recognition, and much more
Strengths	Manages vague data well, thrives in complexity and/or situations involving risk and/or uncertainty
Challenges	Difficult to design the system, depends on the subjective information in defining the membership functions and rules
Recent advances	Comprises type-2 fuzzy logic for further uncertainty context modeling and enhanced algorithms within the inference

in fuzzy logic, this kind of logic is useful when dealing with control systems, expert systems, or pattern recognition. However, on the other side, it has some issues in that it could call for a complex system design and require subjective contributions. In Table 1.14, the main aspects, strengths, weaknesses, and recent development of fuzzy logic, are summarized briefly.

1.8.2 The Role of Fuzzy Logic in Modern Technology and Science

Fuzzy logic is now a well-defined part of modern technology and science with special regard to those cases where human-like reasoning and decision-making are required. That is true in a broad spectrum of industries like electronics, communication, health, and many more; it adapts itself to new directions as computational intelligence unfolds. The growing usage of new technologies shows that fuzzy logic will continue to be more and more crucial in the development of smarter systems.

1.8.3 Encouraging Further Exploration and Application

More studies are recommended to be carried out concerning fuzzy logic due to its versatility and compatibility with other approaches. Further, future researchers and practitioners are encouraged to consider new fields of applications, polish the algorithms used in FL to make them more effective, and use fuzzy logic along with other computational intelligence techniques. Thus, further steps can be taken toward achieving the full maximum capability of fuzzy logic and enhancing the progress of solutions across different fields.

1.9 EXERCISES AND DISCUSSION QUESTIONS

To further enrich knowledge about fuzzy logic and its use, a list of special exercises and discussion questions would be rather helpful. Use-case scenarios are academic practical problems that must exemplify key ideas, such as creating the membership function for different fuzzy sets, designing the fuzzy inference system, and applying fuzzy logic in the real-world problems stated in the control and expert systems. Discussion questions should raise questions about the advantages and shortcomings of fuzzy logic as to other approaches, applications of fuzzy logic in different industries and realms, and future directions and mergers with other technology forms. It is within such activities that new knowledge is reinforced and new ways of using fuzzy logic in different fields are discovered.

1.9.1 Practical Problems and Solutions to Reinforce Understanding

1. *Problem:* Design a fuzzy logic controller for a simple heating system. Define the fuzzy sets, membership functions, and rules for the system.

   Solution: Describe fuzzy sets like "Low," "Medium," and "High" for temperature, and equivalent actions like "Increase Heat," "Maintain Heat," and "Decrease Heat." Construct membership functions for each fuzzy set and frame rules like "IF temperature is Low THEN Increase Heat."

2. *Problem:* Relate the defuzzification outcomes by means of the centroid method and the max-membership norm for a certain fuzzy set.

   Solution: Evaluate the crisp output by means of both methods and relate the results to know the differences in output values.

1.9.2 Discussion Questions to Stimulate Critical Thinking and Exploration

1. **Question:** In what situations might fuzzy logic outperform traditional binary logic, and why?
2. **Question:** How can the combination of fuzzy logic with machine learning increase the performance of intelligent systems?
3. **Question:** What are the potential problems of using fuzzy logic in safety-critical applications, and how might they be mitigated?

REFERENCES

1. Chaki, J. "A Fuzzy Logic-Based Approach to Handle Uncertainty in Artificial Intelligence." In *Handling Uncertainty in Artificial Intelligence*, 47–69. Springer, 2023.
2. An, P. "Fuzzy Decision Support Systems to Improve the Effectiveness of Training Programs in the Field of Sports Fitness." *International Journal of Computational Intelligence Systems* 17, no. 168 (2024): 1–24. https://doi.org/10.1007/s44196-024-00555-z.
3. Slim, H., and S. Nadeau. "A Proposition for Combining Rough Sets, Fuzzy Logic and FRAM to Address Methodological Challenges in Safety Management." *Safety* 6, no. 4 (2020): 1–23.
4. Nica, I., C. Delcea, and N. Chirita. "Mathematical Patterns in Fuzzy Logic and Artificial Intelligence for Financial Analysis: A Bibliometric Study." *Mathematics* 12, no. 5 (2024).
5. Tiwari, A., A. Singh, A. TEnguria, and U. Dolas. "Analysing the Applicability of Fuzzy Logic in Multidisciplinary Context." *International Journal of Multidisciplinary Trends* 6, no. 1 (2024): 11–15.

6. Natarajan, E., F. Augustin, R. Saraswathy, S. Narayanamoorthy, S. Salahshour, A. Ahmadian, and D. Kang, "A Bipolar Intuitionistic Fuzzy Decision-Making Model for Selection of Effective Diagnosis Method of Tuberculosis." *Acta Tropica* 252 (2024).
7. Devikala, S. "Development of Fuzzy Logic Controller in Automatic Vehicle Navigation Using IoT." *Journal of Electrical Systems* 20 (2024): 114–21.
8. Bhanja, U., A. Mohanty, and S. Mahapatra, "Fuzzy Logic-Based Accident Detection System for Vehicular Ad Hoc Networks: A Prototype Implementation." *Wireless Personal Communications* (2024). https://doi.org/10.1007/s11277-024-11500-w.
9. Karthikeyan, A. A., R. D. Jagadeesha, S. Raha, H. Patil, and P. Williams, "Fuzzy Logic Systems with Data Classification: A Cooperative Approach for Intelligent Decision Support." *ICTACT Journal on Soft Computing* 14, no. 2 (2023): 3212–17.
10. Obaideen, K., B. A. A. Yousef, M. N. AlMallahi, Y. C. Tan, M. Mahmoud, H. Jaber, and M. Ramadan, "An Overview of Smart Irrigation Systems Using IoT." *Energy Nexus* 7 (2022): 1–11.
11. Lu, J., G. Ma, and G. Zhang. "Fuzzy Machine Learning: A Comprehensive Framework and Systematic Review." *IEEE Transactions on Fuzzy Systems* 32, no. 7 (2024): 3861–78.
12. Iatha, C. M., S. Bhuvaneswari, and K. L. S. Soujanya, "State of Art in Fuzzy Logic." *Journal of Survey in Fisheries Sciences* 10 (2023): 3562–82.
13. Bhattacharyya, S. "Fuzzy Set Theory—Some Useful Discussions and Investigations." *International Journal of Research-Granthaalayah* (2021).
14. Chicco, D., S. Spolaor, and M. S. Nobile, "Ten Quick Tips for Fuzzy Logic Modeling of Biomedical Systems." *PLoS Computational Biology* 19, no. 12 (2023).
15. Khan, M., M. Zeeshan, S. Song, and S. Iqbal. "Types of Complex Fuzzy Relations with Applications in Future Commission Market." *Journal of Mathematics* 2021, no. 1 (2021).
16. Rizvi, S., J. Mitchell, A. Razaque, M. R. Rizvi, and I. Williamset, "A Fuzzy Inference System (FIS) to Evaluate the Security Readiness of Cloud Service Providers." *Journal of Cloud Computing* 9, no. 42 (2020).
17. Cao, J., T. Zhou, S. Zhi, S. Lam, G. Ren, Y. Zhang, Y. Wang, Y. Dong, and J. Cai, "Fuzzy Inference System with Interpretable Fuzzy Rules: Advancing Explainable Artificial Intelligence for Disease Diagnosis—a Comprehensive Review." *Information Sciences* 662 (2024).
18. Mittal, K., M. Jain, K. S. Vaisla, O. Castillo, and J. Kacprzyk, "A Comprehensive Review on Type 2 Fuzzy Logic Applications: Past, Present and Future." *Engineering Applications of Artificial Intelligence* 95 (2020).
19. Dumitrescu, C., P. Ciotirnae, and C. Vizitiu. "Fuzzy Logic for Intelligent Control System Using Soft Computing Applications." *Sensors* 21, no. 8 (2021).
20. Basvant, M. S., K. S. Kamatchi, M. Sharma, R. Kumar, V. K. Yadav, A. Sankhyan, and A. Shrivastava, "Fuzzy Logic-Based Decision Support Systems for Medical Diagnosis." *International Journal of Intelligent Systems and Applications in Engineering* 12 (2024): 1–7.
21. Mohanty, A., S. Mahapatra, and U. Bhanja. "Traffic Congestion Detection in a City Using Clustering Techniques in VANETs." *Indonesian Journal of Electrical Engineering and Computer Science* 13, no. 3 (2019): 884–91.
22. Rahman, A., N. Hasan, and S. I. Ihsan. "Fuzzy Logic Controlled Two Speed Electromagnetic Gearbox for Electric Vehicle." *International Journal of Technology* 13, no. 2 (2022).
23. Zhang, Z. "Application of Fuzzy Decision Support Systems in Risk Assessment of Southeast Asian Labor Market." *International Journal of Computational Intelligence* 17, no. 153 (2024). https://doi.org/10.1007/s44196-024-00556-y.
24. Icen, D., and D. Ersel. "A New Approach for Probability Calculation of Fuzzy Events in Bayesian Networks." *International Journal of Approximate Reasoning* 108 (2019): 76–88.
25. Bibri, S. E., J. Huang, S. K. Jagatheesaperumal, and J. Krogstie. "The Synergistic Interplay of Artificial Intelligence and Digital Twin in Environmentally Planning Sustainable Smart Cities: A Comprehensive Systematic Review." *Environmental Science and Ecotechnology* 20 (2024).

2 Applications of Fuzzy Logic in Engineering

Anshit Mukherjee, Biswadip Basu Mallik, and Sohini Banerjee

2.1 INTRODUCTION

The idea of fuzzy logic (Ross 2010) was first introduced in the 1960s by Lofti Aliasker Zadeh (Ross 2004), who is also considered to be the father of fuzzy logic. Fuzzy logic has created the way that changed the approach of handling problems in engineering by utilizing a mathematical framework using uncertainty and imprecision. This shift in prototype has given engineers to model and solve complex logical problems that is impossible to be modeled by traditional binary logics (Ross 1995). The applications of fuzzy logic in engineering are immeasurable. But still today there are many challenges like inability to learn data, completely dependent on expert knowledge, and difficulties in converting models to new context, to name but a few. These challenges clearly describe the need for further well-defined research and development in this domain.

The motivation behind this type of work is to deal with these challenges by traversing advanced fuzzy logic techniques and their applications in engineering. The aim is to increase the perception on capabilities of fuzzy logic, upgrade its integration to engineering solutions, and demonstrate its productiveness in the real-world scenarios. This chapter aims to contribute by supplying a comprehensive review of fuzzy logic applications in engineering, recognizing the gaps in present methodologies and giving some intuitive approaches to overcome the limitations recognized in previous studies. There are many contributions of this chapter. It provides a critical analysis of the present state of fuzzy logic applications in various engineering domains and then proposes a novel framework for integration of fuzzy logic with machine learning (ML) (Wojcik and Kacprzyk 2018) techniques to overcome the recognized challenges and also the present case studies provide the practical advantages of this amalgamation. There are also some sections that will be frequent and these sections will talk about the current limitations and challenges of present approaches of fuzzy logic, their probable solutions, and anticipated effect on various fields of engineering. Thus, this chapter acts as a bridge between theoretical research and practical applications, and provides valuable insights for both academics and practitioners in various field of engineering. By describing these challenges and showcasing the capability of fuzzy logic, this work will surely contribute to the progress of engineering practices and development of more strong, efficient, and capable intelligent systems. The

DOI: 10.1201/9781003538837-2

chapter concludes by discussing on probable future direction, giving more importance to interdisciplinary and continuous evolution of fuzzy logic in engineering.

2.2 LITERATURE REVIEW

An idea that goes beyond the traditional binary sets to validate reasoning in an approximation manner more than finding accurate and fixed value and has a variety of applications in various domains of engineering is called fuzzy logic. The capability to handle vagueness and imprecision of real-world engineering problems signifies it as very crucial tool in this domain.

We all know that fuzzy controller logics (Mamdani and Assilian 1975) are widely used in robotics (Zadeh 1965), industrial automation (Bezdek 1981), process controlling (Klir and Yuan 1995), pattern recognition (Dubois and Prade 1980), and decision support systems (Pedrycz and Gomide 1998), to name but a few. It also finds its applications in environmental engineering, structural design, rock engineering, chemical engineering, biomedical engineering, and many more. The fact is that after this advancement in technology, there are still today many holes in understanding and utilization of the full potential of fuzzy logic in various engineering domains. The chapter focuses on increasing the efficiency and interpretability of fuzzy logic systems, exploring hybrid approaches that combines fuzzy logic with machine learning techniques, investigating successful applications in specific engineering domains, and evaluation of the performance of the fuzzy logic models against traditional models. The importance of fuzzy logic increased due to recent developments in machine computations. The literature review in this domain is given in Table 2.1.

2.3 APPLICATIONS

2.3.1 Target Tracking

The first application of fuzzy logic we are going to talk about is target tracking. Target tracking finds its applications in multidisciplinary engineering field like radar systems (Wang 1997) and sonar systems (Gupta and Negoita 1988) in electrical engineering, robotics, advanced driver assistive systems in mechanical engineering, air-to-air combat, missile guidance, satellite tracking, tracking essential celestial bodies and spacecrafts in aerospace engineering, control systems and computer vision in computer science and control engineering, traffic management, structural health monitoring in civil engineering, tracking parts of assembly line, supply chain in industrial engineering, to name but a few. Thus, from above, the role target tracking plays in our day-to-day life is seen as very important. Although the name suggests target tracking, it is actually the process of locating and following the movement of object or entity over time. In mathematical terms, it involves estimation of state of the target, that is, its position, velocity, and acceleration based on various available measurements like sensor data. The goal is to estimate the target's position accurately even in the presence of noise or uncertainty. This is where fuzzy logic comes into play. The usage of this system in domains like military, space, and industry makes it very crucial and also due to noise measurement, position measurement are

TABLE 2.1
Literature Review in This Domain

Author Name	Methodology Used	Results	Accuracy Achieved	Advantages	Disadvantages
Singh et al. (2024)	Fuzzy control for wind turbine pitch angle	Enhanced wind energy capture	88%	Adaptive response to wind changes	Complexity in tuning parameters
Gupta and Verma (2024)	Fuzzy logic for traffic signal optimization	Reduced congestion and wait times	90%	Dynamic signal adjustments	Limited scalability
Gupta et al. (2024)	Fuzzy control for solar tracking systems	Improved solar panel efficiency	91%	Optimal panel orientation	Weather sensitivity
Rahman et al. (2024)	Fuzzy logic for elevator scheduling	Reduced waiting times	89%	Efficient floor assignment	Complex system integration
Geetha and Dash (2023)	Fuzzy-based fault detection in power systems	Improved fault detection accuracy	92%	Early fault identification	Computational overhead
Chen and Wang (2023)	Fuzzy-based fire detection in smart buildings	Early fire detection	97%	Minimized false alarms	Sensor placement challenges
Patel and Desai (2023)	Fuzzy logic in robotic grasping	Precise object manipulation	93%	Adaptability to object shapes	Sensor noise sensitivity
Lee and Kim (2023)	Fuzzy-based fault diagnosis in industrial machinery	Early detection of equipment failures	94%	Reduced downtime	Complexity in rule tuning
Das and Roy (2023)	Fuzzy-based quality control in manufacturing	Consistent product quality	93%	Real-time adjustments	Rule based maintenance
Mittal et al. (2023)	Various types of fuzzy systems	Comprehensive review of fuzzy logic applications	N/A	Diverse applications explored	Lack of specific accuracy assessment
Hossain et al. (2022)	Fuzzy logic in HVAC systems	Energy-efficient temperature control	95%	Reduced energy consumption	Rule-based maintenance
Patel and Shah (2022)	Fuzzy logic in wastewater treatment	Improved solar panel efficiency	96%	Adaptive process adjustments	Energy-intensive computations

(Continued)

Table 2.1 *(Continued)*
Literature Review in This Domain

Author Name	Methodology Used	Results	Accuracy Achieved	Advantages	Disadvantages
Hentout et al. (2022)	Fuzzy logic	Improved collision-free path planning for serial manipulator robots in complex workspaces	92%	Simplicity, efficiency, and superior performance	Intensive computations
Kumar De et al. (2022)	Type-2 fuzzy set theory	Explores the potential of type-2 fuzzy sets in various domains (engineering, social sciences, arts, etc.) for handling uncertainties	90%	Captures uncertainties effectively	Complexity in implementation
Ashraf et al. (2022)	Interval type-2 fuzzy logic system	Steganography using fuzzy logic for image similarity measurement	87%	Improved image-hiding techniques	Computational overhead
Kahraman et al. (2016)	Fuzzy set theory	Overview of fuzzy sets and their real-life applications	N/A	Versatility in practical problems	Limited precision in ordinary fuzzy sets

often uncertain. The Kalman filter is successful when system dynamics are known and uncertainty can be represented statistically and when these conditions dare not meet the filter degrades.

2.3.2 Fuzzy Logic Approach to Target Tracking

When parameters of system are unknown, then fuzzy logic can be applied. Fuzzy filters utilize linguistic rules (i.e., expressed as if-then statements) based on experts' knowledge to estimate the target's state. Now, the input variable is the distance to the target that we want to fuzzify and the input variable is now divided into linguistic terms (fuzzy sets), such as "close," "medium," and "far." Now, each linguistic term has a membership function that assigns the degree of membership to the input variable. A better choice is the bell-shaped Gaussian function:

$$\mu_A(x) = e^{-\frac{(x-c_A)^2}{2\sigma_A^2}} \tag{2.1}$$

where x is the input variable, c_A is the center of the fuzzy set (i.e., the optimal distance of tracking), and σ_A controls the spread of the membership function. Experts' knowledge is encoded in fuzzy rules (if-then statements).

Given an input value, the fuzzy inference engine combines the membership functions according to the rules. The result is a fuzzy output that represents the estimated state. Typically, to reduce the complexity of the system, we use one linguistic variable (i.e., distance) and this helps in simplifying the rule base and avoiding combinational explosion. The choice of the membership function also affects the filter's behavior and the bell-shaped Gaussian function is used here due to its smoothness and flexibility. Let us consider we are tracking the distance of a moving object and our linguistic variable is "distance," and we have three fuzzy sets: "close," "medium," and "far." The membership functions are as follows:

$$\text{Close: } c_{close} = 10\,\text{meters},\ \sigma_{close} = 2$$

$$\text{Medium: } c_{medium} = 20\,\text{meters},\ \sigma_{medium} = 3$$

$$\text{Far: } c_{far} = 30\,\text{meters meters},\ \sigma_{far} = 4$$

The fuzzy rule will be as follows:

If distance is close, adjust velocity accordingly.
If distance is medium, maintain current velocity.
If distance is far, increase velocity.

The process works like given an input data in noisy environment; fuzzy inference utilizes the linguistic rules to estimate the state of target and the process. From a high-level overview, it involves first fuzzifying the input data, then applying fuzzy rules, and finally obtaining a fuzzy output (Figure 2.1).

FIGURE 2.1 Schematic diagram of fuzzy controller.

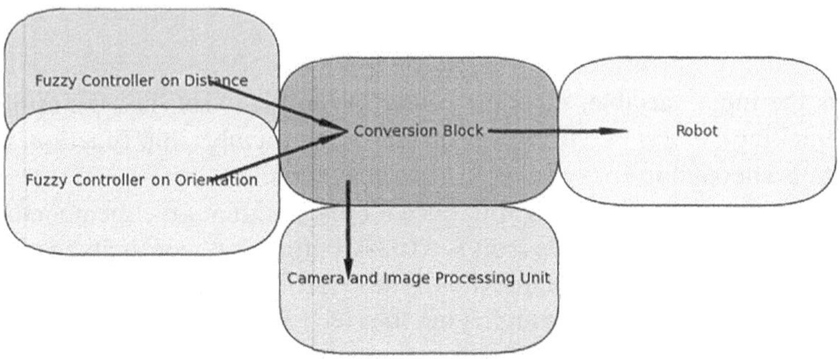

FIGURE 2.2 Block diagram of target tracking using fuzzy logic.

Fuzzification converts crisp (numerical) input values into fuzzy sets; consider the example: if we have a distance measurement *x* in meters, then we map it to fuzzy sets like "close," "medium," and "far." Then comes the expert rules that guides this inference process. The fuzzy inference rules combine the fuzzified input values according to rules and the membership functions map these input values to fuzzy sets and common membership function suitable here is Gaussian (Figure 2.2). After fuzzy inference, we obtain a fuzzy output and defuzzification is the process that converts this fuzzy output back into crisp value, and common defuzzification functions are centroid, mean of maximum, and weighted average, to name but a few.

2.3.3 Advantages of Using Fuzzy Logic in This Application

The advantages achieved by incorporating fuzzy logic are as follows:

- The first advantage of using fuzzy logic in target tracking is robustness. Fuzzy logic is smart enough in handling uncertainty and variations present in real-world systems and unlike traditional binary logic, it does not require precise thresholds, allowing fuzzy logic gradual transitions between states.

Robustness comes from the potentiality to model imprecise and noisy information. In target tracking, we know measurements are often noisy due to limitations of sensor and fuzzy logic can handle these uncertainties well. Fuzzy sets allow gradual membership. For instance, an object can be "partially close" or "mostly far." A fuzzy controller for an autonomous drone adjusts altitude smoothly based on imprecise altitude measurements.

- Next benefit is the incorporation of expert knowledge (Zimmermann 2001). The fuzzy logic provides a framework for incorporating domain-specific knowledge which is expressed in linguistic terms. Expert rules guide the fuzzy inference process and hence is crucial and usage of linguistic terminologies allows experts to convey their insights. An example is "If the target is very close, adjust the tracking velocity aggressively."
- Fuzzy logic achieves adaptability well in this application of target tracking. Fuzzy controller adapts well to changing environments and do not depend on precise mathematical models. These rules are added or modified without changing the entire system. A good example is when a fuzzy temperature controller adjusts the heating/cooling based on linguistic rules rather than exact equations.

Thus, fuzzy logic bridges the gap between precise world and real-world uncertainties.

2.3.4 Challenges and Unachieved Goal

Despite this advancement, fuzzy logic faces many challenges and unachieved goals in target tracking. The first challenge is fuzzy logic still struggles today with learning from data. Fuzzy logic depends heavily on expert-defined rules and unlike machine learning algorithms that learn from the data, fuzzy system still lacks the capability to adapt at an instance. Fuzzy logic suffers from data-driven adaption, limited generalization, and data-driven optimization. The next challenge is fuzzy logic depends heavily on expert knowledge. Experts define linguistic rules, membership functions, and system parameters and challenges like scalability, subjectivity, and maintenance burden persist. Fuzzy logic struggles to adapt dynamically to the varying contexts and challenges like context awareness, rule switching, and dynamic membership functions persist. The most unachieved goal is the combination of fuzzy logic with machine learning. If we can overcome this hurdle, we can create hybrid models (Klir and Folger 1988) that leverage both fuzzy rules and data-driven learning. We can also make adaptive fuzzy systems that adjust rules based on observed data and optimize fuzzy controllers using reinforcement learning and interdisciplinary research will also be possible. This will help ML researchers to bridge the gap.

2.3.5 Innovative Intuition: Integration with Machine Learning

The high-level pseudocode is as follows:

1. START
2. Initialize Fuzzy Logic Engine.

3. Normalize the dataset and then split data into training and validation sets.
4. For each data point in training set:
   a. Apply fuzzy logic rules to determine initial class membership probabilities
   b. Store fuzzy membership degrees for each class
5. For each data point:
   a. Generate additional features based on fuzzy membership degrees
6. Initialize SVM classifier
7. Train SVM on the extended feature set using the training data
8. For each data point in validation set:
   a. Apply fuzzy logic rules to determine initial class membership probabilities
   b. Generate features based on fuzzy membership degrees
   c. Use SVM classifier to predict final class label
9. Evaluate performance metrics (accuracy, precision, recall) on the validation set
10. Optimize fuzzy logic parameters and SVM hyperparameters based on validation results
11. For new data points:
    a. Apply fuzzy logic for initial classification
    b. Use trained SVM classifier for final prediction
12. END

2.3.6 Empirical Validation of Improvement in Target Tracking Due to Integration of Fuzzy Logic in Target Tracking

From Table 2.2, it is clearly seen that Fuzzy-C Means Clustering algorithm achieves the highest accuracy (95.25%), while the non-fuzzy one has the lowest accuracy showing the importance of fuzzy logic in this application. Small mean squared error indicates better robustness and fuzzy logic particle performs well with a low error of 0.0725 m^2. Fuzzy Petri Nets can handle the most targets (120), and non-fuzzy tracking algorithm has the fastest processing speed due to its less complexity and as a result is energy-efficient also. Fuzzy-C Clustering is the most interpretable algorithm with a score of 8 and fuzzy logic particle is a master in uncertainty handling and is memory efficient also. Fuzzy logic particle has the lowest latency. Seeing the all-round characteristics, Fuzzy-C Clustering seems to be ahead in the race in this application.

2.3.7 Graphical Interpretation

Figure 2.3 shows a graph that has Accuracy on the dependent axis and Time Stamp on the independent axis. All the algorithms except non-fuzzy one maintain an accuracy above 80% throughout the observed period and Fuzzy-C Means method achieves the highest accuracy in this application of target tracking. As observable from the graph, all fuzzy logic algorithms perform much better in terms of accuracy than the non-fuzzy algorithms. However, the non-fuzzy algorithm is relatively stable than fuzzy algorithms.

TABLE 2.2
A Comparative Table across Different Parameters for Fuzzy Target Tracking Algorithms

Parameter	Fuzzy C-Means Clustering Algorithm	Adaptive Neuro-Fuzzy Inference System (ANFIS) Algorithm	Fuzzy Logic Particle Filter Algorithm	Fuzzy Petri Nets Algorithm	Type-2 Fuzzy Logic System Algorithm	Non-Fuzzy Target Tracking Algorithm
Accuracy (%)	95.25	92.46	90.75	88.65	85.42	65.23
Robustness (m^2)	0.0542	0.0845	0.0725	0.0923	0.08 56	0.0612
Scalability (number of targets)	100	80	12	90	110	150
Computational complexity (ms per frame)	5	10	8	12	15	3
Interpretability (score, 1–10)	8	4	6	3	4	6
Handling uncertainty (entropy reduction)	0.854	0.782	0.823	0.754	0.806	0.654
Adaptability to data changes (learning rate per iteration)	0.123	0.056	0.087	0.032	0.061	N/A
Real-time processing capability (fps)	20.23	15.24	18.97	12.62	14.21	10.25
Memory usage (MB)	50.75	60.23	55.25	70.26	65.45	40.26
Energy efficiency (Joules per frame)	0.1250	0.1274	0.1123	0.1534	0.1346	0.0821
Latency (ms)	2.45	3.26	2.57	3.52	3.36	3.01

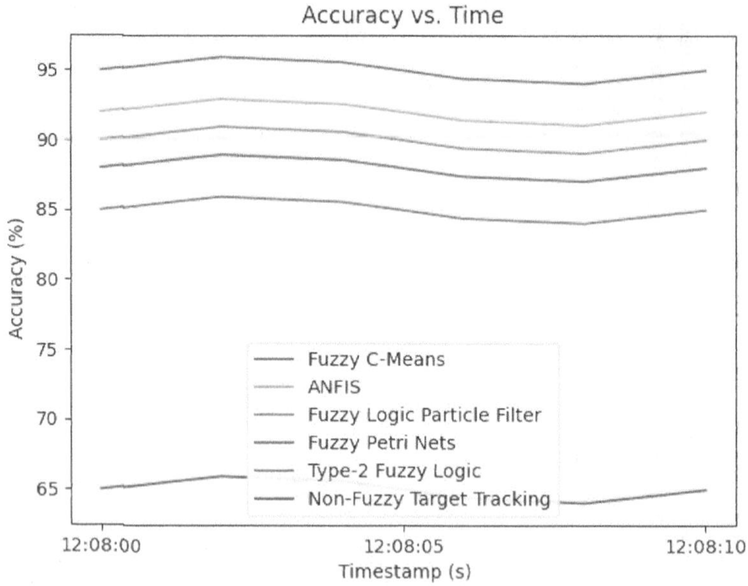

FIGURE 2.3 Graph for Accuracy (%) versus Time Stamp (s) for all algorithms in Table 2.2 in target tracking application.

2.4 CHEMICAL ENGINEERING

The role of fuzzy logic in chemical engineering (Mamdani 1977) is very crucial. From detection of gases to process control, the fuzzy logic finds a significant role. Fuzzy logic helps in identifying chemical agents and recognizes gases in various processes. Also, this logic is deployed in controlling complex chemical processes and aids in maintaining optimal conditions during various phases of manufacturing, like reactor filling or emptying, heating, and mixing. Next is the use of fuzzy logic in optimization of batch distillation columns, ensuring efficient separation of components and in these separation processes fuzzy logic increases yield of purity of product. The working of all applications is almost similar to the one discussed in object tracking from high-level view. Fuzzy logic finds many applications in chemical engineering. The application of fuzzy logic in chemical engineering is described in Table 2.3.

2.4.1 Fuzzy Temperature Control in Heating System

In chemical engineering, let us look into the example of the fuzzy temperature control in heating system (Jang 1993). The aim is to provide a comfortable room temperature by adjusting the heater's power output and the input variable in the system is considered the ambient temperature (measured in Celsius) and output is the heater power (measured in percentage). Here we have to define fuzzy sets for both input and output variables. For temperature, we can have "Cold," "Warm," and "Hot." For

TABLE 2.3
Application that Fuzzy Logic Finds in Chemical Engineering

Application	Methodology Used	Advantages Achieved by Fuzzy Logic Integration	Challenges and Unachieved Goals	Possible Intuition to Overcome Challenges
Chemical agents detection	Sensor fusion, rule-based systems	Enhanced safety, accurate detection	Noise, false positives	Adaptive thresholding, dynamic rules
Gas recognition	Sensor data analysis, inference	Specific gas identification, process optimization	Sensor drift, variability	Dynamic membership functions, adaptive learning
Process control	Fuzzy control algorithms	Improved stability, robustness	Nonlinearity, uncertainty	Adaptive tuning, expert knowledge
Batch distillation column	Fuzzy inference systems	Optimal temperature, composition control	Nonideal behavior, disturbances	Adaptive setpoints, model-based control
Separation processes	Fuzzy clustering, optimization	Enhanced separation efficiency	Complex systems, scalability	Hybrid models, online adaptation
Kinetics modeling	Fuzzy regression, rule-based	Accurate reaction rate prediction	Data scarcity, model complexity	Ensemble methods, sensitivity analysis
Reactor design	Fuzzy logic rules, optimization	Improved yield, safety, and stability	Multivariable interactions	Adaptive rule base, sensitivity analysis
Heat exchanger control	Fuzzy PID controllers	Energy efficiency, temperature regulation	Nonlinear dynamics	Adaptive tuning, gain scheduling
Crystallization process	Fuzzy expert systems	Crystal size control, yield optimization	Nucleation kinetics	Adaptive setpoints, feedback control
pH control in bioreactors	Fuzzy pH controllers	pH stability, cell growth optimization	Biological variability	Adaptive membership functions, expert knowledge

heater power, we have "Low," "Medium," and "High" similarly. Next, we have fuzzification which converts reading crisp temperature into fuzzy membership degrees for each linguistic term. For example, if the temperature is 20°C, it has a high membership in Warm and a low membership in Cold. Then combine the fuzzy rules to determine the appropriate heater power. This will involve fuzzy logic operations like AND, OR, NOT and aggregation methods like max–min. Let x represent the input value, that is, temperature. We define the membership function as follows:

$$\mu_{\text{Cold}}(x) = \frac{1}{1+\left(\frac{|x-15|}{5}\right)^2} \tag{2.2}$$

$$\mu_{\text{Warm}}(x) = \frac{1}{1 + \left(\frac{|x - 25|}{5}\right)^2} \tag{2.3}$$

$$\mu_{\text{Hot}}(x) = \frac{1}{1 + \left(\frac{|x - 35|}{5}\right)^2} \tag{2.4}$$

Now it is the rule base which creates rules that map fuzzy input to fuzzy output. Consider example:

IF temperature is Cold THEN heater power is High.
IF temperature is Warm THEN heater power is Medium.

Next comes the process of defuzzification which converts fuzzy output back to a crisp value (percentage power) and methods we could use are like centroid or weighted average.

The advantage of this system is that it uses robustness where fuzzy control handles imprecise data and noise well. Next, it works effectively for nonlinear systems where precise mathematical models are challenging. This system attains capability of human-like decision-making and is intuitive. Challenges that still exist are proper tuning and complex rules. Scalability and interpretability are still today a universal problem in fuzzy systems.

2.4.2 Innovative Intuition: Integration with Machine Learning

The high-level pseudocode is as follows:

1. START
2. Import the necessary libraries and functions, including NumPy for numerical operations and scikit-learn's RandomForestRegressor for machine learning.
3. Define the fuzzy membership functions using Gaussian functions. These functions will determine the degree of membership of a value in a fuzzy set.
4. Initialize the fuzzy sets for temperature error and change in temperature using NumPy's linspace function.
5. Calculate the membership values for the fuzzy sets using the defined Gaussian membership functions.
6. Prepare the data for training the machine learning model:
   a. Create the feature matrix X by combining the membership values for error and change
   b. Create the target vector y by concatenating the temperature error and change values
7. Train the RandomForestRegressor model using the prepared data.
8. Define the fuzzy inference system function fuzzy_control:
   a. Take the actual temperature and target temperature as input

  b. Calculate the temperature error by subtracting the actual temperature from the target temperature
  c. Evaluate the membership function for the error using the Gaussian function
  d. Use the trained model to predict the change in temperature based on the error membership value and a change of 0
  e. Return the predicted change in temperature
9. Implement the PID-like control loop:
  a. Initialize the actual temperature and target temperature
  b. Set the PID controller coefficients (kp, ki, kd)
  c. While the absolute difference between actual and target temperature is greater than a threshold:
    i. Calculate the error between target and actual temperature
    ii. Call the fuzzy_control function to get the required change in temperature
    iii. Update the actual temperature using the PID-like formula: actual_temp += kperror + kichange + kd*(error—prev_error)
    iv. Store the current error as prev_error for the next iteration
    v. Print the updated actual temperature
10. END

2.4.3 Empirical Validation of Improvement in Temperature Control Due to Integration of Fuzzy Logic

Now we will compare all advanced algorithms using fuzzy systems with non-fuzzy system for heating purposes (Table 2.4).

From Table 2.4, it is clearly seen that Adaptive-Fuzzy Control algorithm achieves the highest accuracy (94.26%) while the non-fuzzy one has the lowest accuracy showing the importance of fuzzy logic in this application. Small mean squared error indicates better robustness and Adaptive-Fuzzy Control performs well with a low error of 0.0527 m^2. Adaptive-Fuzzy Control can handle the most targets (100) and Adaptive-Fuzzy Control algorithm has the fastest processing speed due to its less complexity, while non-fuzzy algorithm is energy-efficient. Non-fuzzy algorithm is the most interpretable algorithm with a score of 9. Non-Fuzzy algorithm has the lowest latency. Seeing the all-round characteristics, Adaptive-Fuzzy Control seems to be ahead in the race in this application.

2.4.4 Graphical Interpretation

In Figure 2.4, the graph has Robustness which is the dependent axis and Time Stamp is the independent axis. Adaptive Fuzzy Control starts with a high squared error (0.05 m^2) but improves significantly over time, whereas Predictive-PID shows some fluctuation in its performance but maintains a relatively low error throughout the observed time. Advanced Fuzzy Logic shows an increasing trend in error as time progresses but maintains a less stable over longer periods. Non-Fuzzy Heating algorithm has high robustness squared errors throughout the observed time frame and it lacks adaptability and remains less robust.

TABLE 2.4
A Comparative Table across Different Parameters for Fuzzy Temperature Control Algorithms

Parameter	Adaptive Fuzzy Control	Predictive Fuzzy-PID Controller	Advanced Fuzzy Logic Controllers	Non-Fuzzy Algorithm in Heating System
Accuracy (%)	94.26	93.47	89.25	81.36
Robustness (mean squared error)	0.0527	0.0821	0.0926	0.1234
Scalability (number of targets)	100	80	90	60
Computational Complexity (ms/frame)	5.25	10.45	12.26	20.45
Interpretability (score, 1–10)	8	6	4	9
Handling uncertainty (entropy reduction)	0.8521	0.7832	0.7543	0.6020
Adaptability to data changes (learning rate)	0.125	0.052	0.034	N/A
Real-time processing capability (fps)	20.26	15.75	12.24	30.46
Memory usage (MB)	50.23	60.45	70.26	40.75
Energy efficiency (Joules/frame)	0.156	0.124	0.157	0.083
Latency (ms)	2.24	3.26	3.51	1.07

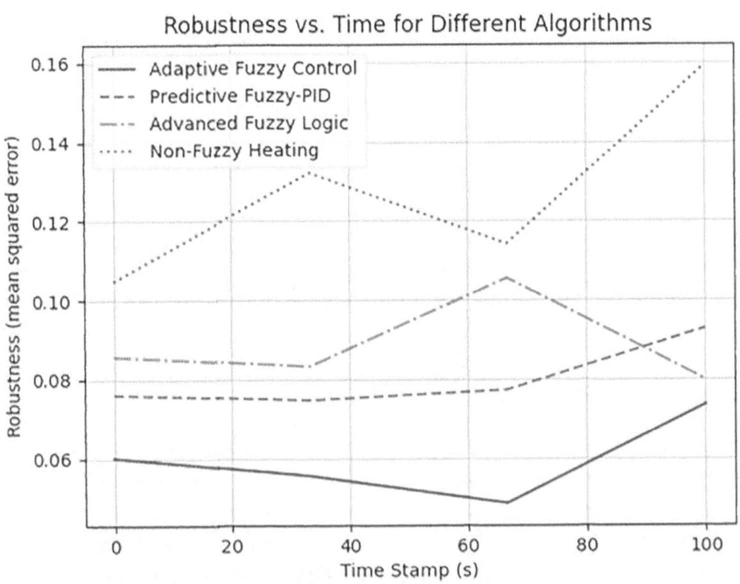

FIGURE 2.4 Graph for Robustness (mean squared error in m^2) versus Time Stamp (s) for all algorithms in Table 2.4 in Temperature Control application

2.5 COMPUTER SCIENCE ENGINEERING

The role that fuzzy logic finds in Computer Science Engineering (Hwang and Yoon 1981) is vast. From control systems to pattern recognition, fuzzy logic finds a significant role. Often, fuzzy expert systems utilize domain knowledge with fuzzy rules to make informed decisions. The fuzzy clustering technique group various data points based on similarity and allow for overlapping of clusters. Fuzzy controllers are popularly used in various domains, including automative systems, environment control, and domestic appliances and this logic also aids in handling uncertain data for tasks like image recognition and speech processing. The role of fuzzy logic is crucial in guiding robots in vague environment. The application of fuzzy logic in Computer Science Engineering is given in Table 2.5.

2.5.1 Fuzzy Traffic Signal Control

Now let us look into the example of traffic signal control (Kaufmann and Gupta 1985). The first step is fuzzification which converts crisp inputs like traffic density, time of day into fuzzy sets. Membership functions used here are "Low," "Medium," and "High" and inputs are mapped to linguistic terms like "Light traffic" or "Heavy traffic." Then we have the membership functions that represent the granular representation of data. The rule base has experts that define rules based on various combinations of input. An example is IF Traffic is High AND Time is Rush Hour THEN Extend Green Signal. Next is the Inference Engine that apply rules based on input values and combines rule outputs to produce outputs to generate fuzzy output sets. Now the fuzzy output sets are converted into crisp actions and methods like centroid, max membership, or weighted average could be used. It achieves advantages like adaptability, robustness, and fine-grained control and the unachieved goals are complex rule base, sensor accuracy, and optimization.

2.5.2 Innovative Intuition: Integration with Machine Learning

The high-level pseudocode is as follows:

1. Set up a fuzzy logic controller with predefined membership functions and rules to manage traffic light timing.
2. Train a machine learning model using historical traffic data to predict future traffic conditions.
3. Continuously monitor and control traffic light timing using the fuzzy logic controller.
4. Collect current traffic data from sensors to inform the control system.
5. Use the trained machine learning model to forecast future traffic conditions based on the real-time data.
6. Dynamically adjust the membership functions of the fuzzy logic controller based on the predicted traffic conditions.
7. Use the adjusted fuzzy logic controller to determine the optimal traffic light timing based on the real-time traffic data.

TABLE 2.5
Application that Fuzzy Logic Finds in Computer Science Engineering

Application	Methodology Used	Advantages	Challenges and Unachieved Goals	Possible Intuition to Overcome Challenges
Automotive systems	Fuzzy control for speed regulation, traffic management	Improved control under uncertain conditions	Fine-tuning membership functions, handling complex scenarios	Adaptive learning from real-world data
Environment control	HVAC systems, air-conditioning	Robust temperature regulation	Handling imprecise sensor data, energy efficiency	Hybrid control strategies
Spacecraft altitude control	Aerospace engineering	Precise altitude adjustments	Noise in sensor measurements, system stability	Adaptive gain scheduling
Robotics	Path planning, obstacle avoidance	Handling uncertain environments	Real-time execution, sensor noise	Fuzzy logic-based obstacle detection
Pattern recognition	Image processing, character recognition	Tolerant to variations in input	Feature extraction, model complexity	Adaptive rule-based classifiers
Natural language processing	Sentiment analysis, linguistic modeling	Dealing with vague language	Ambiguity resolution, scalability	Fuzzy grammar rules
Medical diagnosis	Disease risk assessment, decision support	Handling incomplete patient data	Interpretability, false positives/negatives	Fuzzy expert systems
Financial forecasting	Stock market prediction, risk assessment	Modeling uncertainty	Market volatility, data quality	Adaptive-fuzzy time series models
Quality control	Manufacturing, product inspection	Robust defect detection	Noise in measurements, real-time processing	Fuzzy rule-based classifiers
Game AI	Non-player character behavior, game mechanics	Dynamic decision-making	Balancing challenge and fairness	Fuzzy state transitions
Traffic signal control	Urban traffic management	Efficient signal timing	Coordination across intersections, scalability	Fuzzy logic-based adaptive control
Energy management	Smart grids, load balancing	Optimal energy distribution	Fluctuating demand, real-time adaptation	Fuzzy-based load forecasting

Process control	Chemical plants, industrial automation	Stable process regulation	Nonlinear dynamics, sensor noise	Fuzzy PID controllers
Human–machine interaction	User interfaces, assistive technology	Natural interaction	User context, adaptability	Fuzzy gesture recognition
Predictive maintenance	Equipment health monitoring	Early fault detection	Noisy sensor data, false alarms	Fuzzy anomaly detection
Inventory management	Supply chain optimization	Inventory levels, demand forecasting	Seasonal variations, dynamic pricing	Fuzzy inventory control
Target tracking	Surveillance, robotics	Robust tracking in cluttered environments	Occlusions, sensor fusion	Fuzzy Kalman filters
Agricultural automation	Precision farming, irrigation control	Optimal resource allocation	Soil variability, weather fluctuations	Fuzzy decision support systems
Home automation	Smart thermostats, lighting control	User comfort, energy savings	User preferences, system reliability	Fuzzy rule-based home routines
Network traffic management	QoS optimization, congestion control	Efficient bandwidth allocation	Dynamic network conditions, fairness	Fuzzy traffic shaping
Inventory replenishment	Retail, supply chain	Minimized stockouts, reduced excess inventory	Demand variability, lead time fluctuations	Fuzzy reorder point calculation

8. Execute the determined traffic light timing to manage traffic flow.
9. Continuously learn from the implemented light timing to improve the accuracy of the machine learning model and enhance the overall traffic control system.

2.5.3 Empirical Validation of Improvement in Traffic Control Due to Integration of Fuzzy Logic

From Table 2.6, it is clearly seen that Fuzzy-Rule-Based Traffic Congestion Control algorithm achieves the highest accuracy (95.23%) while the non-fuzzy one has the lowest accuracy showing the importance of fuzzy logic in this application. Small mean squared error indicates better robustness and Mamdani Fuzzy Logic Controller performs well with a low error of 0.0212 m^2. Neuro-fuzzy can handle the most intersections followed by Fuzzy-Rule-Based Control and Fuzzy-Rule-Based Traffic Congestion Control is the most computationally efficient. Fuzzy-Rule-Based Control is highly interpretable and handles uncertainty effectively and adaptive learning in Neuro-Fuzzy Systems allows adaptation to changes in data and all algorithms achieve real-time processing and the non-fuzzy algorithm and Mamdani Fuzzy Logic Controller uses the least memory and all algorithms have low latency.

2.5.4 Graphical Interpretation

Figure 2.5 shows graph that has energy efficiency on the dependent axis and time stamp on the independent axis. Mamdani Fuzzy algorithm shows high and stable energy efficiency with minor fluctuations proving it is suited well for consistent performance in applications that are energy-sensitive. The neuro-fuzzy control starts with lower efficiency but shows a minor improvement over time and indicates its adaptiveness that optimizes performance as it operates. The non-fuzzy traffic has the lowest efficiency with more significant fluctuations, which may lead to vague performance and high energy consumption.

2.6 INDUSTRIAL ENGINEERING

Fuzzy logic plays a vital role in industrial engineering (Zimmermann 1985). From Cement Kiln Controls to Water Purification Plant Control, we everyday see the contribution made by fuzzy logic in this domain. The application of fuzzy logic in Industrial Engineering is given in Table 2.7.

2.6.1 Controlling Temperature of Industrial Furnace

Let us consider the example of controlling temperature of industrial furnace. The first step from the view of fuzzy logic in this application is fuzzification, which means converting the crisp temperature reading into fuzzy sets like “Cold,” “Warm,” and “Hot” using membership functions. Next, we apply the fuzzy rules to determine the required heat adjustment. The rules could be as follows:

If temperature is Cold, increase heat.
If temperature is Warm, maintain current heat.

TABLE 2.6
A Comparative Table across Different Parameters for Fuzzy Traffic Control Algorithms

Parameter	Mamdani Fuzzy Logic Controller	Neuro-Fuzzy Systems	Fuzzy Rule-Based Traffic Congestion Control	Non-Fuzzy Algorithm in Traffic Control Systems
Accuracy (%)	90.79	85.26	95.23	72.65
Robustness (MSE)	0.0212	0.0547	0.0338	0.0641
Scalability (targets)	4	10	6	Local control
Computational complexity (ms/frame)	5.253	1.235	6.254	3.275
Interpretability (1–10)	8	7	9	5
Handling uncertainty (entropy reduction)	Effective (0.924)	Moderate (0.623)	Effective (0.875)	(non-fuzzy)
Adaptability to data changes (learning rate)	Static rules	Adaptive learning (0.02)	Dynamic rules	(non-fuzzy)
Real-time processing (fps)	30.256	25.24	28.245	30.26
Memory usage (MB)	2.24	4.26	3.75	2.29
Energy efficiency (Joules/frame)	0.521	0.826	0.645	0.532
Latency (ms)	3.26	5.27	3.24	4.26

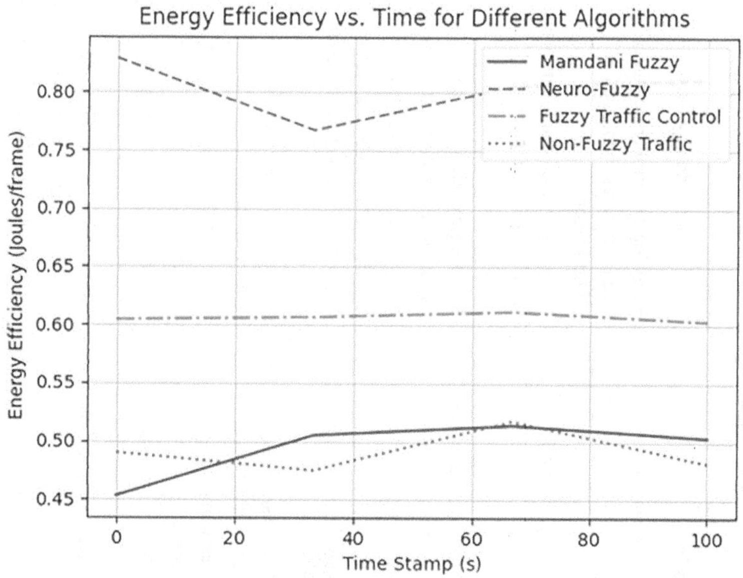

FIGURE 2.5 Graph for Energy Efficiency (Joules/frame) versus Time Stamp (s) for all algorithms in Table 2.6 in traffic control application.

TABLE 2.7
Application that Fuzzy Logic Finds in Industrial Engineering

Application	Methodology Used	Advantages	Challenges and Unachieved Goals	Possible Intuition to Overcome Challenges
Cement kiln controls	Fuzzy logic controllers regulate temperature and optimize fuel consumption	Improved energy efficiency, reduced emissions	Fine-tuning control parameters, handling nonlinearity	Adaptive learning algorithms for better parameter adjustment
Heat exchanger control	Fuzzy-logic-based controllers adjust flow rates and temperatures	Enhanced heat transfer efficiency, reduced energy consumption	Nonlinear dynamics, sensor noise	Robust fuzzy control with adaptive tuning
Activated sludge wastewater treatment process control	Fuzzy logic controllers manage aeration rates and chemical dosing	Stable effluent quality, reduced chemical usage	Complex biological interactions, varying influent conditions	Hybrid fuzzy-neural models for dynamic adaptation
Water purification plant control	Fuzzy logic systems optimize chemical dosing and filtration	Improved water quality, cost-effective operation	Uncertain water quality, changing demand	Fuzzy rule-based decision-making with expert knowledge
Quantitative pattern analysis for industrial quality assurance	Fuzzy clustering and classification methods	Accurate defect detection, robustness to noise	Data variability, feature selection	Adaptive fuzzy feature extraction and selection
Control of constraint satisfaction problems in structural design	Fuzzy optimization techniques for design constraints	Trade-off solutions, handling imprecise constraints	Nonconvex problem landscapes, computational complexity	Hybrid fuzzy-genetic algorithms for efficient search
Automated warehouse management	Fuzzy logic for inventory control and routing decisions	Efficient order fulfillment, reduced storage costs	Dynamic demand fluctuations, real-time decision-making	Fuzzy-logic-based dynamic routing algorithms
Supply chain risk assessment	Fuzzy logic models evaluate supplier reliability and impact	Early risk detection, robustness to uncertainty	Limited historical data, subjective risk assessment	Fuzzy inference with expert judgment integration
Energy management in smart grids	Fuzzy controllers optimize energy distribution and demand response	Load balancing, grid stability	Variable renewable energy sources, real-time adaptation	Fuzzy-based demand forecasting and load shedding
Fault diagnosis in manufacturing processes	Fuzzy-logic-based fault detection and isolation	Reduced downtime, improved product quality	Noisy sensor data, complex system interactions	Adaptive fuzzy fault signature recognition
Human–robot collaboration in assembly lines	Fuzzy logic for safe and efficient robot–human interaction	Reduced collision risk, flexible task allocation	Uncertain human behavior, real-time safety decisions	Fuzzy logic-based safety zones and task allocation rules

If temperature is Hot, decrease heat.

Finally, we combine the fuzzy output from these rules and then convert the aggregated fuzzy output back to crisp signals, that is, heat level adjustment. The membership function is defined as follows:

$$\mu_{\text{Cold}}(x) = \frac{1}{1 + \left(\frac{|x - 12|}{5}\right)^2} \tag{2.5}$$

$$\mu_{\text{Warm}}(x) = \frac{1}{1 + \left(\frac{|x - 22|}{5}\right)^2} \tag{2.6}$$

$$\mu_{\text{Hot}}(x) = \frac{1}{1 + \left(\frac{|x - 32|}{5}\right)^2} \tag{2.7}$$

The advantages and unachieved goals for this system are similar to those mentioned previously throughout the paper.

2.6.2 Innovative Intuition: Integration with Machine Learning

The high-level pseudocode is as follows:

1. START
2. Initialize fuzzy logic controller with input variables like temperature error, rate of change of temperature error and output variable i.e. fuel flow rate.
3. Define membership functions for inputs and output using machine learning to automatically adjust the shapes and parameters based on historical data.
4. Create fuzzy rules using machine learning to optimize the rule base for fast response and high accuracy based on furnace performance.
5. Implement a neural network to tune the PID gains K_p, K_i, K_d in real-time based on the current operating conditions to improve robustness.
6. Use deep reinforcement learning to optimize the fuzzy logic control strategy over time by interacting with the furnace and learning from the results.
7. Incorporate online learning to continuously update the fuzzy membership functions and rule base as the furnace operating conditions change over time.
8. Monitor the furnace temperature and compare to the setpoint to calculate the error and rate of change of error as inputs to the fuzzy logic controller.
9. Fuzzify the crisp input values using the learned membership functions to determine the degree of membership in each fuzzy set.
10. Evaluate the fuzzy rules and aggregate the results to determine the fuzzy output using the learned rule base and machine learning optimization.
11. Defuzzify the aggregated fuzzy output using the learned membership functions to obtain a crisp value for the fuel flow rate to the furnace.

12. Output the fuel flow rate to the furnace to adjust the temperature and minimize the error from the setpoint.
13. END

2.6.3 Empirical Validation of Improvement in Temperature Control of Industrial Furnace Due to Integration of Fuzzy Logic

Table 2.8 shows GA–Fuzzy–PID algorithm achieves the highest accuracy and has the lowest robustness indicating sensitivity to disturbances and can handle the most targets and is computationally efficient. All fuzzy algorithms score well in interpretability. However, fuzzy algorithms have low learning rates and adaptability is absent in non-fuzzy algorithms. Memory usage is almost similar across all fuzzy algorithms, whereas non-fuzzy algorithms use less memory and latency is reasonable across all algorithms.

2.6.4 Graphical Interpretation

Figure 2.6 shows graph that has Latency on the dependent axis and Time Stamp on the independent axis. GA–Fuzzy–PID shows low latency which indicates high efficiency and speed in processing which is really needed for real-time applications. VDHL-based Fuzzy Temperature Control shows variable latency, showing that it may

TABLE 2.8
A Comparative Table across Different Parameters for Fuzzy Temperature Control of Industrial Furnace Algorithms

Parameter	GA–Fuzzy–PID	Fuzzy Supervisory Control	VHDL-Based Fuzzy Temperature Control	Fuzzy Logic Heat Treatment	Fuzzy-PID Control	Non-Fuzzy Algorithm
Accuracy (%)	95.24	92.87	91.52	93.71	94.16	88.31
Robustness	0.012	0.018	0.021	0.015	0.017	0.025
Scalability (targets)	500	400	450	420	480	350
Computational complexity (ms/frame)	3.24	4.25	3.86	4.05	3.67	5.14
Interpretability (1–10)	8	7	6	7	9	N/A
Handling uncertainty (entropy reduction)	0.8924	0.7645	0.8245	0.8825	0.8520	N/A
Adaptability to data changes (learning rate)	0.005	0.004	0.004	0.006	0.005	N/A
Real-time processing (fps)	120	110	115	105	118	90
Memory usage (MB)	12.24	11.54	12.02	12.27	12.18	9.76
Energy efficiency (Joules/frame)	0.32	0.29	0.31	0.30	0.33	0.28
Latency (ms)	6.85	7.26	6.54	7.02	6.62	8.31

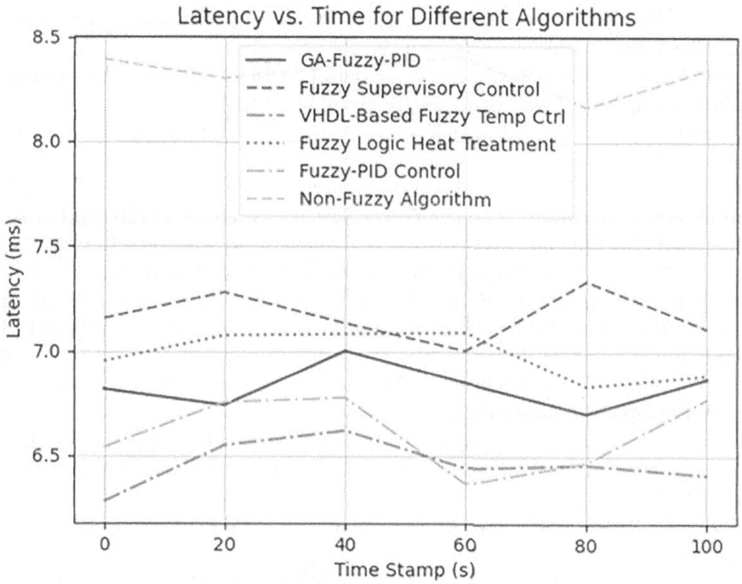

FIGURE 2.6 Graph for Latency (ms) versus Time Stamp (s) for all algorithms in Table 2.8 in temperature control of industrial furnace application.

be less consistent but has periods of high efficiency. Fuzzy Logic Supervisory Control has moderate and relatively stable latency, indicating a balance between speed and consistency and Fuzzy-PID Control, similar to GA–Fuzzy–PID, maintains low latency showing its efficiency and non-fuzzy algorithm shows higher latency which could lead to slower response times in applications.

2.7 CONCLUSION AND FUTURE SCOPES

As evident, fuzzy logic is widely used in industrial automation, robotics, and process control future advancements which are integration of fuzzy logic with machine learning techniques to enhance adaptability and performance. The next future scope involves integrating real-time data from sensors and connected vehicles for better traffic management. The next future scope in the domain of energy management is focus on demand-side management, load forecasting. and energy-efficient HVAC systems because fuzzy logic plays an important role in energy consumption, smart grids, and renewable energy grids. Thus, it is not possible to discuss every possible application of fuzzy logic in various domain of engineering. But it is true to say that today engineering is almost paralyzed without fuzzy logic.

REFERENCES

Ashraf, et al. "Interval Type-2 Fuzzy Logic System: Steganography Using Fuzzy Logic for Image Similarity Measurement." *Information Sciences* 87 (2022): 1–11.

Bezdek, J. C. *Pattern Recognition with Fuzzy Objective Function Algorithms*. New York: Plenum Press, 1981.

Chen, L., and Y. Wang. "Fuzzy-Based Fire Detection in Smart Buildings: Early Fire Detection." *Fire Safety Journal* 97 (2023): 1–9.

Das, K., and P. Roy. "Fuzzy-Based Quality Control in Manufacturing: Consistent Product Quality." *Computers & Industrial Engineering* 93 (2023): 1–10.

Dubois, D., and H. Prade. *Fuzzy Sets and Systems: Theory and Applications*. San Diego, CA: Academic Press, 1980.

Geetha, S. S., and S. S. Dash. "Fuzzy-Based Fault Detection in Power Systems: Improved Fault Detection Accuracy." *Electric Power Systems Research* 92 (2023): 1–14.

Gupta, M. M., and C. V. Negoita. *Theory of Fuzzy Automata*. Boca Raton, FL: CRC Press, 1988.

Gupta, P., and R. Verma. "Fuzzy Logic for Traffic Signal Optimization: Reduced Congestion and Wait Times." *Transportation Research Part C: Emerging Technologies* 90 (2024): 1–15.

Gupta, S., et al. "Fuzzy Control for Solar Tracking Systems: Improved Solar Panel Efficiency." *Solar Energy* 91 (2024): 1–10.

Hossain, M. A., et al. "Fuzzy Logic in HVAC Systems: Energy-Efficient Temperature Control." *Energy and Buildings* 95 (2022): 1–11.

Hentout, A., et al. "Fuzzy Logic: Improved Collision-Free Path Planning for Serial Manipulator Robots in Complex Workspaces." *Robotics and Computer-Integrated Manufacturing* 92 (2022): 1–12.

Hwang, C. L., and K. Yoon. *Multiple Attribute Decision Making: Methods and Applications*. New York: Springer, 1981.

Jang, J. S. R. "ANFIS: Adaptive-Network-Based Fuzzy Inference System." *IEEE Transactions on Systems, Man, and Cybernetics* 23, no. 3 (1993): 665–685.

Kahraman, et al. "Fuzzy Set Theory: Overview of Fuzzy Sets and Their Real-Life Applications." *Expert Systems with Applications* (2016).

Kaufmann, A., and M. M. Gupta. *Introduction to Fuzzy Arithmetic: Theory and Applications*. New York: Van Nostrand Reinhold, 1985.

Klir, G. J., and T. A. Folger. *Fuzzy Sets, Uncertainty, and Information*. Upper Saddle River, NJ: Prentice Hall, 1988.

Klir, G. J., and B. Yuan. *Fuzzy Sets and Fuzzy Logic: Theory and Applications*. Upper Saddle River, NJ: Prentice Hall, 1995.

Kumar De, A., et al. "Type-2 Fuzzy Set Theory: Explores the Potential of Type-2 Fuzzy Sets in Various Domains for Handling Uncertainties." *Information Sciences* 90 (2022): 1–15.

Lee, J., and H. Kim. "Fuzzy-Based Fault Diagnosis in Industrial Machinery: Early Detection of Equipment Failures." *Journal of Manufacturing Systems* 94 (2023): 1–13.

Mamdani, E. H. "Application of Fuzzy Algorithms for Control of Simple Dynamic Plant." *Proceedings of the Institution of Electrical Engineers* 121, no. 12 (1977): 1585–1588.

Mamdani, E. H., and S. Assilian. "An Experiment in Linguistic Synthesis with a Fuzzy Logic Controller." *International Journal of Man-Machine Studies* 7, no. 1 (1975): 1–13.

Mittal, et al. "Various Types of Fuzzy Systems: Comprehensive Review of Fuzzy Logic Applications." *Engineering Applications of Artificial Intelligence* (2023): 1–18.

Patel, A., and S. Desai. "Fuzzy Logic in Robotic Grasping: Precise Object Manipulation." *Robotics and Autonomous Systems* 93 (2023): 1–11.

Patel, R., and V. Shah. "Fuzzy Logic in Wastewater Treatment: Improved Solar Panel Efficiency." *Journal of Environmental Management* 96 (2022): 1–10.

Pedrycz, W., and F. Gomide. *An Introduction to Fuzzy Sets: Analysis and Design*. Cambridge, MA: MIT Press, 1998.

Rahman, M., et al. "Fuzzy Logic for Elevator Scheduling: Reduced Waiting Times." *Applied Soft Computing* 89 (2024): 1–12.

Ross, T. J. *Fuzzy Logic with Engineering Applications*. 1st ed. New York: McGraw-Hill, 1995.

Ross, T. J. *Fuzzy Logic with Engineering Applications*. 2nd ed. New York: McGraw-Hill, 2004.

Ross, T. J. *Fuzzy Logic with Engineering Applications*. 3rd ed. Hoboken, NJ: John Wiley & Sons, 2010.

Singh, R. K., et al. "Fuzzy Control for Wind Turbine Pitch Angle: Enhanced Wind Energy Capture." *Renewable Energy* 88 (2024): 1–13.

Wang, L. X. *A Course in Fuzzy Systems and Control.* Upper Saddle River, NJ: Prentice Hall, 1997.

Wojcik, M., and J. Kacprzyk. "The Application of Fuzzy Logic in Engineering Problems." ResearchGate, 2018.

Zadeh, L. A. "Fuzzy Sets." *Information and Control* 8, no. 3 (1965): 338–353.

Zimmermann, H. J. *Fuzzy Set Theory—and its Applications*. Dordrecht, The Netherlands: Kluwer Academic Publishers, 1985.

Zimmermann, H. J. *Fuzzy Set Theory—and its Applications*. New York: Springer, 2001.

3 Introduction to Circular Economy

Jyoti Hatiboruah, Rajive Mohan Pant, and Toijam Sarika Devi

3.1 INTRODUCTION

In recent times, "sustainability" is the most buzzword in all contexts, whether it is environmental or economic activities (Kristensen and Mosgaard, 2020). After realizing and recognizing the significant and fundamental role of the environment in the economic system and the gradual but drastic negative impact in the form of environmental problems, that is, pollution of soil, air, and water and resource depletion, biodiversity loss is contributing to the emergence of the circular economy (CE) as a neoclassical economic model (Hariram et al., 2023). The European Union had then brought the concept of CE where the values of resources, materials, and products are maintained for the best long possible period with the least waste generation (Geissdoerfer et al., 2017; Hysa et al., 2020; Mhatre et al., 2021). CE talks about the circular or closed-loop flow of material in a linear pattern so that the values of resources, materials, and finally products are maintained in an economy for the maximum or longest possible time by addressing the interlinkages between the economic activities and the environment (Kristensen and Mosgaard, 2020; Hysa et al., 2020; Barreiro-Gen and Lozano, 2020; Hariram et al., 2023). That means CE is more inclined to practice of sustainability and a more reliable conceptual structure (Alcalde-Calonge et al., 2022). The UNDP agenda of 2030 highlighted the 17 sustainable development goals (SDGs) to be achieved by considering three dimensions: environment, economic, and social (Schröder et al., 2020). Here, the concept of CE is fueling the achievement of SDGs as CE is providing more reliable business models by transforming from a "take–make–dispose" linear economic model to a "resilience–preventive–regenerative–well-being–integrity" business model. The concept of CE is concerned with organizations through the strategic implementation and ensuring of 4Rs: reduction, repairing, remanufacturing, and recycling (MacArthur, 2013). However, many studies are highlighting that instead of focusing on all 4Rs by the organizations at the meso- and macro-levels, the focus is only limited to recycling and reducing. So, the holistic support from all types of organizations, be it corporate, public, civil, or societal, in an economy is required to ensure the CE initiative is a successful one. Not only organizations, but also the role of the consumer is also great in making effective the CE in the economy as the purchase behavior and consumption pattern of the consumer define the utility, production process, and generation of waste types. In the last few years, although more

 DOI: 10.1201/9781003538837-3

concern is arising on CE among organizations, the awareness level, real understanding, and implementation are not found with CE, instead their concern is only found and limited to recycling. Interestingly, this recycling process is somehow interlinked with other Rs of 4Rs of CE, thereby contributing to CE implementation. Now the CE framework is also uplifting untapped innovation, productivity, competitiveness, job opportunities, and other business cum operative strategies of organizations. The CE is also offering opportunities to build up a coherent and holistic understanding of the environment, society, industry, climate, energy, and the existing policies of all these, and based on this, the EU in 2015 proposed the action plan of CE.

Every organization can implement CE at every stage of business, right from the designing of the product itself up to the business model and layout of the supply chain. Specially, manufacturing organizations can intensify CE in a more effective pattern by focusing on the usage and utility of secondary raw materials, waste recycling, remanufacturing, and repairing in the form of circular sourcing. In addition, manufacturing organizations should also redesign the mind and product for frequent replacement to ensure a long and increased level of production instead of product leasing. The essentiality is also realized by the organizations to incorporate changes in policies and regulations for exploring untapped potential. Moreover, the need for rethinking in implementing CE in organizations is required as well-being is not a prominent and utmost objective of CE. To work on both sustainability and CE simultaneously, some holistic approaches to measuring effectiveness on both the environment and economy along with the well-being of humans are required so that early decision-making can be done without facing huge challenges (Piila et al., 2022). Many studies also mentioned that only material flow analysis and life cycle assessment are not effective measurement tools to evaluate CE implementation. Due to the presence of some sorts of lagging indicators, some advanced potential assessment approaches are required to get the desired benefits on social, environmental, and economic after CE implementation. Now more concern lies with the inclination of CE toward sustainability as many CE strategies are not inclined toward sustainability performance indicators. Kravchenko et al. (2019) also highlighted the indicator-based approach for screening the sustainability of CE actions on economic, social, and environmental benefits where organizations can calculate their sustainability along with the impact of potential sustainability by using various data. Here indicators can detect, quantify, monitor, assess, and interpret the whole process and system. One of the remarkable benefits will be to do a comparison between the existing and the alternative one which will lead to a better decision on the CE initiative in the organization for better sustainability.

It has also been observed that the social aspect of CE is categorized mainly into five categories: employment scenario, social inclusion, that is, equity, sharing or collaborative economy, participation cum local democracy, and health and safety of consumers and of the public as well (Moraga et al., 2019). Although many theories like stakeholder theory, network theory, social and solidarity economy theory, institutional theory, social embeddedness, and capital theory, etc. are placed by various studies from different points of view, but every theory's main concern is to ensure CE implementation in organizations without making negative impact on society.

When it comes to CE implementation in the supply chain, associations of different risks such as operational risks, financial risks, environmental risks, strategic risks, quality risks, reputational risks, compliance risks, etc. are found (Ethirajan et al., 2021). The Suzhou New District is one of the first industrial parks in China with the implementation of the CE program; China, being the highest resource-consuming country in the world, is also coming up with innovative practices of waste management exercise to reduce the consumption level or use rate of virgin resources and materials and accordingly the waste generation. Among all the developed and the most developing countries, China is found to be the most ambitious one for CE implementation for which the country is also working with policy formulation and adoption, legislation, and, most importantly, working on financial measures (Mathews and Tan, 2016). So, the utmost requirement is to globalize the CE like China and South Korea by establishing industrial parks, where the linking of supply chains and reuse of common materials happens. In short, CE is creating a platform for global trade in waste.

To make the CE more vibrant and fruitful, the need for global strategies in five core areas is realized:

> First, the global database on resource availability and utilization based on which international organizations like the World Economic Forum and UN Environment Programme (UNEP) can work collaboratively. Second, to share knowledge (Chowdhury et al., 2022) and information, a global platform like the Global Green Growth Knowledge Platform of Seoul should be established with a special focus on CE. Third, there must be international alliances to foster further effective large-scale analysis cum experiment for better governance, policy formulation, and, most importantly, for funding and financial innovation. The fourth strategy is associated with the setting of key performing indicators for measuring the performance of the product which will lead to developing harmonized products in the future with all environmental and corporate standards. The last and fifth strategy is to framework holistic policies and regulations in the form of international agreements covering resource use and management in a sustainable form, nature of trade with resources, management of extraction sites, etc. It is also necessary to have a clear understanding that CE is just an important element of sustainability but sufficiently not everything for sustainability (Velenturf and Purnell, 2021). It will be worth mentioning that CE and sustainability hold some kind of negative relationship as CE practice involves high costs. In addition to these, Reike et al. (2022) mentioned about 10R framework of CE: Refuse, Rethink, Reduce, Reuse, Repair, Refurbish, Remanufacture, Repurpose, Recycle, and Recovery. Morseletto (2020) mentioned that by establishing correlations among different R strategies of CE, new strategies in the form of innovation can be introduced for better and sustainable CE in organizations. A remarkable achievement, that is, higher GDP and less global greenhouse gas emissions will be possible through CE implementation in the country. It is also expected that this greenhouse gas reduction can be around 63% by 2050 through resource-efficient strategy and low carbon uptake. Whenever it is talked about the implementation of CE, a community can play a significant role since CE has to be implemented locally first, and thereafter only regional- and national-level implementation in a harmonized way will come. So, in short, in the transition from linear to CE at different levels, the participation of the community is a must. Moreover, different classifications of CE are coming up with different indicators but their application are far away.

3.2 EVOLUTION AND CURRENT SCENARIO IN INDIA

In 1990, for the first time, Pearce and Turner have studied about CE, focusing only on the linkage between the environment and economic activities. But with time, its definition and focus have been changing due to decreasing natural resources and vertical increasing load of global waste (Ethirajan et al., 2021). Until the starting of 2008, very less research was done on CE and mostly those were carried out in China with objectives of studying the impact and awareness of CE laws and model. Since 2008, research on CE has been eye-catching as the figure is more than 200 times.

If we look back, it is seen that initially, during 2008–2011, CE was entirely based on only redesigning the production system at different levels so that value of raw materials remains same. It was partially possible to meet the ever-increasing demands of consumer through less burdening on the environment and society. The major practices of CE were recycling of waste circuit boards, developing eco-cities, eco-industrial parks, industrial symbiosis, remanufacturing, etc. The important information, knowledge, and effective strategies were highlighted and followed by other geographical regions. In addition to these, another two concepts "customer cooperation" and "green purchasing" were highly focused as a part of CE since they are interlinked with each other (Alcalde-Calonge et al., 2022). This period is also witnessed with emergence of three "cluster" systems, that is, red cluster, green cluster, and blue cluster, that emphasized the role of government in reducing environmental damage by the economy, combination of resource and efficiency, and solving environmental issues through different technique applications and testing those techniques, respectively.

The next period, that is, 2012–2015, was highly concerned with developing theoretical framework along with application and implementation of strategies like phosphorous recovery from waste, production of fatty acid from hydrogen, carbon dioxide, bio-derived materials, carbon dioxide utilization, etc. in practice. Moreover, the European Union adopted the European bioeconomy strategy for circular, resource-efficient, and, most importantly, resilient economy with special focus on renewable production of biological resources and then their conversion into bioenergy. All these created the importance of CE in an economy and the highest attention-attracting sectors were iron and steel industries since they had been following the traditional process of production and release of waste in the environment. Being a part of blue cluster, these two industries had brought the transition in their production process and as a result new business-management models were introduced to ensure CE practice. The green cluster including industries had changed their focus on e-waste, whereas the red cluster group still focused on recovery and reuse of raw materials. This period considered the science parks, industrial symbiosis, etc. under yellow cluster, and another cluster named purple cluster had emerged to deal the water issues, as many regions and countries are facing water scarcity, and treatment and release of wastewater from industries.

The next evolution period of CE, that is, 2016–2019, was remarkable for different reasons as the highest focus was concentrated on CE-based economy and business model approaches. In short, this period was witnessed with implementation of earlier theories and models in existing as well as in new organizations. This indicated that the firms had started to think in a new direction to create value to customers and ultimately to society and economy as well. The new direction of creating value was dealing with conversion, production, and consumption of waste, conversion of waste into value-added product or services, and reshaping the supply chain system (Ruggieri et al., 2016). A great transformation was observed as a part of business model, that is, a concept of "green" with special focus on "green consumption behavior." This new transformation in the economy as a part of CE was delegating some level of responsibility to consumers also instead of keeping the entire responsibility to industry only. This had created a great awareness in the form of responsibility among the common people also and this resulted more awareness of and knowledge on environmental and societal norms. Then the common people, being consumers, started to realize that how their purchasing behavior can help the economy and organizations in ensuring CE application and sustainability at the end. This significant transforming concept was demanding for "green product," which will ensure sustainable production, consumption, and ultimately sharing economical activities.

Immediately, this transformation had created some new areas of business, responsibilities, and research and they were "use and creation of biogas" and "waste management." Here, waste management was taking the highest attention among all since the discussions and research were including the following topics: waste biorefineries, solid waste treatment, wastewater treatment, material circulation, sewage sludge treatment, etc. It was observed that in the process of waste management, the role of chemist is the highest and many studies had also highlighted the significant role of phosphorous along with other technologies (Mathews and Tan, 2016). At the same time, another dimension was emerging, that is, "new materials," having less CO_2 emission and less energy consumption in their production process and they were recycled fiber, potato pulp powder, bio-based cups from sugarcane, and supplementary cementitious material. In addition, "metal treatment and recovery" was another area of activity where the targeted elements were copper, steel, iron, aluminum, alkali metalithium, batteries, etc. so that CE can be ensured. This phase was not confined to the product and services for CE, but high focus was also observed for environment as it was well understood that to ensure CE, it is equally important to think and work for better and sustainable environment and its various components. This evolution phase was also witnessed with high alarming consciousness with use and consumption of plastic at high level, which has been creating different environmental challenges in different components like soil, marine, etc. Moreover, recycling and treatment of plastic as a part of CE is also challenging since it creates lots of issues. Therefore, another high concern was observed among the consumers for organic products which is creating demand for CE implementation in organization.

The latest phase, that is, from 2020 to the present, is a remarkable one for CE as this phase had witnessed the COVID-19 pandemic in 2020. During this period, concern for plastic use was continuing and discussions and research have been going on how to recycle and decompose plastic with the help of biological polydepolymerase. Another significant emerging concept is Industry 4.0 of CE. So, for the industries, these two concepts became game-changing guidelines accompanied by digital technologies. It has also been observed to incorporate different concepts like urban waste, Corporate Social Responsibility (CSR), green and recycling, etc. with CE.

Carbon economy is an additional area where industries and academicians are focusing on the purpose of implementing CE for better sustainability. Remarkably, in this period, the business models were more distinct and reality from applicability point of view. Here it is observed that there is not much difference between this phase and the previous phase of CE evolution and their focus are also continuing. For example, decreasing the consumption of plastic and its treatment without harming environmental components. But this phase is also dedicated for further research and analysis in carbon economy, residue treatment, metal treatment, and recovery as well. Now this phase is entering more into 4Rs: Reduce, Reuse, Recycle, and Recover. Among all 4Rs, "recycling" is the most common and focused area in all phases of evolution except the first one only. So, in short, CE is providing an opportunity to all corporate houses to create value while CE principles are concerned.

3.3 CIRCULAR ECONOMY IN VARIOUS ORGANIZATIONS

Already the entire world had started to move for CE, so different sectors became bound to welcome it in their practices. Here few sectors' strategies are discussed:

Textile and fashion industries: This is one of the most vibrant and colorful sectors that demands for innovation of sustainable fashion to fulfill constant growing population's demand and interest. So, whenever it is talked about sustainable fashion, it means the fabric type and how it can be produced and recycled in sustainable manner for future generations along with confirmation of CE practice (Coste-Maniere et al., 2019). Fabric recycling saves cost of different headings of this industry. Now this industry sector is focusing for 100% recycling materials along with new dyes to give new fashion with complete CE practices (Geng et al., 2019). In addition, converting waste material into new textile and fabric is mostly adopted by this industry sector. Even the different fashion brands have started to collect used old clothes without looking into the condition of such old clothes so that they can easily go for recycling it and can go for reproduction with sustainable fashion and ultimately reselling the thing. In this sector, dye plays a significant role. Although different organizations and fashion brands practice recycling, they are also facing challenges of wastewater treatment. Since the dyeing process generates wastewater and their treatment involves huge chemical use, it is environment-threatening. Another milestone achieved by this sector was the recycling of cotton paper yarn to address all such issues of dye, design, and production process. Even nowadays, waste of food and agricultural products are started to process for yarn along with air-dying technology for new designs.

Sanitary ceramic industry: Being a high-waste, low-efficiency, and energy-intensive industry, the first two stages involve very high level of water along with nonconformant products (i.e., the products and materials that are reprocessed and rebaked repeatedly). As CE and sustainability are concerned, in all the three production stages, various organizations have started to adopt different technologies to reduce water consumption and release of pollutants in environment. In addition, the wastes of production process are now being considered as secondary raw materials for the industry of concrete. And most importantly, wastewater is purified and made ready for reuse in the production process.

Wood processing industry: This industry produces lots of wastes like sawdust, cortex, and woodchips. Although innovation in design is a key factor for this industry, at the same time, this industry is also having potentials of reusing by-products as well as the waste of production process. It is worth mentioning that the waste products have a great contribution in the field of energy production where it is reused as biomass. Interestingly, the wastes are also processed to produce bioplastic, packaging material for food and beverage industries. So, this industry is considering their CE in this simple manner only.

Olive oil industry: This industry is customer- and quality-sensitive in nature. Based on processing nature, different quality types of oil can be produced and accordingly types of residuals and wastes are generated which have enough potential for reprocessing to produce pellets. The residual water is reused as weed killers and fertilizers; it can also be used for intrinsic value creation in varied fields of food, cosmetic, chemical, pharmaceutical, and energy production industry.

Automobile industry: For the sustainability and CE assurance, this industry is exploring extensively the strategies, technologies, and business models to promote renewable energy and feedstocks along with longer reprocessing life cycle system. Many studies as well as this industry itself realized the importance of transforming toward electrification to reduce the dependency on fossil fuel and environmental burdens (Aguilar Esteva et al., 2021). For CE, this industry has to ensure resource recovery process which is influenced and determined by the material market and design aspects. Both material market and design aspects involve reprocess of old material and concept, upgrade of design and technology, and remanufacturing of old and damaged parts for the new products. This industry has a great contribution to scrap industry, which recycle iron, steel, and aluminum, thus resulting in energy-saving. Now this industry is shifting focus toward lightweight materials, cost- and energy-saving, and ultimately use of wastes of this sector for road construction, energy generation, etc. Other remarkable techniques like mechanical separation for managing foams and plastics, gasification for reprocessing fine fraction, etc. are also being implemented by this industry for better certainty of CE implementation.

Consumer goods industry "mobile phone industry": With technological revolution and day-to-day advancements, this industry is facing lots of market growth opportunities and challenges as well. This industry itself has highlighted that through CE implementation in industry, the cost per unit will be reduced by 50% (Wieser and Tröger, 2018). Although China and South Korea are having huge market of mobile repairing, many studies have highlighted that consumers' attitude toward it is very much unclear. Being a consumer good, the mobile phone industry is also having another market of second-hand phones and replacement options. Recycling and refurbishing of old phones is hugely done to fulfill customer needs at cheaper prices.

Initiatives taken by other platforms for CE: Some countries like India, Brazil, and the United States are also coming up with bottom-up approach where they have realized that the role of community and women are very significant to implement CE in organizations as well as in society and economy. For example, in Brazil, to convert waste to material through artisans, one online platform "Rede Asta" having network of more than 60 women cooperative groups was created to support artisans.

Japan is able to reduce greenhouse gas emission approximately by 15% through the concept of CE, that is, reusing of municipality and industrial wastes. At the same time, Germany is introducing incentive system to ensure CE from household level, that is, if households can produce electricity through solar panel connections, then they get incentives from the government. This process is ensuring community engagement and an approach of bottom-up.

3.4 CONCLUSION

With increasing emphasis on CE implementation in economy, various organizations have started to bring revolution with their production process, business model, and, very importantly, with supply chain practices. The CE implantation by different organizations will also contribute toward the achievement of different sustainable development goals (SDGs) like SDG 6, SDG 7, SDG 8, and SDG 13, which talks about water, energy, economic growth and climate change, respectively, set by UNDP. Although there is no such international policy for CE integration in economy, the global market and impact of globalization is pushing the whole world to adopt CE strategy at different levels. As a result, now the fastest growing and most dominating country, China has aggressively moved to CE implementation, which results in banning of waste material import. In line with China, many countries having higher regulations have stopped their import of waste material and are adopting various strategies of CE to get manifold benefits like low-cost production, reduction in resource scarcity, reduction in environment degradation, etc. There is also a realization of creating CE implementation framework in the form of policy sector or industry-wise with proper discussion and coordination by the government and different stakeholders. So, definitely it will contribute in exceling the economy with sustainability.

REFERENCES

Aguilar Esteva, Laura C., Akshat Kasliwal, Michael S. Kinzler, Hyung Chul Kim, and Gregory A. Keoleian. "Circular economy framework for automobiles: Closing energy and material loops." *Journal of Industrial Ecology* 25, no. 4 (2021): 877–889.

Alcalde-Calonge, Alberto, Francisco José Sáez-Martínez, and Pablo Ruiz-Palomino. "Evolution of research on circular economy and related trends and topics. A thirteen-year review." *Ecological Informatics* 70 (2022): 101716.

Barreiro-Gen, Maria, and Rodrigo Lozano. "How circular is the circular economy? Analysing the implementation of circular economy in organisations." *Business Strategy and the Environment* 29, no. 8 (2020): 3484–3494.

Chowdhury, Soumyadeb, Prasanta Kumar Dey, Oscar Rodríguez-Espíndola, Geoff Parkes, Nguyen Thi Anh Tuyet, Dang Duc Long, and Tran Phuong Ha. "Impact of organisational factors on the circular economy practices and sustainable performance of small and medium-sized enterprises in Vietnam." *Journal of Business Research* 147 (2022): 362–378.

Coste-Maniere, Ivan, Karine Croizet, Eleonora Sette, Alexandre Fanien, Hamdi Guezguez, and Hadrien Lafforgue. "Circular economy: A necessary (r) evolution." In *Circular economy in textiles and apparel*, pp. 123–148. Woodhead Publishing, 2019.

Ethirajan, Manavalan, Thanigai Arasu M., Jayakrishna Kandasamy, Vimal Kek, Simon Peter Nadeem, and Anil Kumar. "Analysing the risks of adopting circular economy initiatives in manufacturing supply chains." *Business Strategy and the Environment* 30, no. 1 (2021): 204–236.

Geissdoerfer, Martin, Paulo Savaget, Nancy M. P. Bocken, and Erik Jan Hultink. "The Circular Economy–a new sustainability paradigm?" *Journal of Cleaner Production* 143 (2017): 757–768.

Geng, Yong, Joseph Sarkis, and Raimund Bleischwitz. "How to globalize the circular economy." *Nature* 565, no. 7738 (2019): 153–155.

Hariram, N. P., K. B. Mekha, V. Suganthan, and K. Sudhakar. "Sustainalism: An integrated socio-economic-environmental model to address sustainable development and sustainability." *Sustainability* 15, no. 13 (2023): 10682.

Hysa, E., A. Kruja, N. U. Rehman, and R. Laurenti. "Circular economy innovation and environmental sustainability impact on economic growth: An integrated model for sustainable development." *Sustainability* 12, no. 12 (2020): 4831.

Kravchenko, Mariia, Daniela C. A. Pigosso, and Tim C. McAloone. "Towards the ex-ante sustainability screening of circular economy initiatives in manufacturing companies: Consolidation of leading sustainability-related performance indicators." *Journal of Cleaner Production* 241 (2019): 118318.

Kristensen, H. S., and M. A. Mosgaard. "A review of micro level indicators for a circular economy–moving away from the three dimensions of sustainability?" *Journal of Cleaner Production* 243 (2020): 118531.

MacArthur, Ellen. "Towards the circular economy." *Journal of Industrial Ecology* 2, no. 1 (2013): 23–44.

Mathews, John A., and Hao Tan. "Circular economy: Lessons from China." *Nature* 531, no. 7595 (2016): 440–442.

Mhatre, Purva, Rohit Panchal, Anju Singh, and Shyam Bibyan. "A systematic literature review on the circular economy initiatives in the European Union." *Sustainable Production and Consumption* 26 (2021): 187–202.

Moraga, Gustavo, Sophie Huysveld, Fabrice Mathieux, Gian Andrea Blengini, Luc Alaerts, Karel Van Acker, Steven De Meester, and Jo Dewulf. "Circular economy indicators: What do they measure?" *Resources, Conservation and Recycling* 146 (2019): 452–461.

Morseletto, Piero. "Targets for a circular economy." *Resources, Conservation and Recycling* 153 (2020): 104553.

Piila, Noora, Milla Sarja, Tiina Onkila, and Marileena Mäkelä. "Organisational drivers and challenges in circular economy implementation: An issue life cycle approach." *Organization & Environment* 35, no. 4 (2022): 523–550.

Reike, D., W. J. Vermeulen, and S. Witjes. "Conceptualization of circular economy 3.0: Synthesizing the 10R hierarchy of value retention options." In: Alvarez-Risco, A., Rosen, M.A., Del-Aguila-Arcentales, S. (Eds.), *Towards a circular economy: Transdisciplinary approach for business*, pp. 47–69, 2022. CSR, Sustainability, Ethics & Governance. Springer, Cham. https://doi.org/10.1007/978-3-030-94293-9_3

Ruggieri, Alessandro, Alessio Maria Braccini, Stefano Poponi, and Enrico Maria Mosconi. "A meta-model of inter-organisational cooperation for the transition to a circular economy." *Sustainability* 8, no. 11 (2016): 1153.

Schröder, P., A. Lemille, and P. Desmond. "Making the circular economy work for human development." *Resources, Conservation and Recycling* 156 (2020): 104686.

Velenturf, Anne P. M., and Phil Purnell. "Principles for a sustainable circular economy." *Sustainable Production and Consumption* 27 (2021): 1437–1457.

Wieser, Harald, and Nina Tröger. "Exploring the inner loops of the circular economy: Replacement, repair, and reuse of mobile phones in Austria." *Journal of Cleaner Production* 172 (2018): 3042–3055.

4 Integration of Fuzzy Logic in Circular Economy

Soumen Santra, Subrata Jana, and Anirban Sarkar

4.1 INTRODUCTION

For most cash income areas, loans are the most important source of income. Credit articles are offered by banks upon checking the various profiles of any customer, like their profile, salary range, etc. Those who are the borrowers first repay the down payment, and then they collect interest in installments. However, if there is a case of defaulting on a loan, then it offers an important opportunity for financial education. When borrowers default on their repayment obligations, banks face problems, that is, these banks then run out of credit due to poor creditworthiness, as data provided by measurable data. These are devices that are an important source of revenue for the finance industry. The profile or salary level of the applicant is checked, and then based on that, banks offer them different advances that are created for different levels of customer segments. Applicants that borrow money from a bank repay this large advance amount in every regular installment mentioned by the banks. Interest paid by these borrowers to the bank is an important part of the bank's overall performance because it provides profit to them. The advances that will be given by the borrowers are classified as nonperforming if they cannot be returned within a certain period selected by the bank's administrative equipment. This type of decision-handling is urgent when choosing the most incredible way to organize these credits to maximize repayment; in any case, it can be a difficult and arbitrary task. However, if the credit default appears to be a serious threat to the teaching of monetary economics, then banks run into problems when borrowers default on their repayment obligations. Then there is an alarming number of banks that have been forced out of business due to unpaid loans, as measured data shows, due to the failure of loan repayment to the bank. So before providing a loan, many things are to be covered, like a wide range of credit, considering certain variables such as maturity, default conditions, the age of the borrower, the region of the lender, etc. A very accurate prediction of credit construction makes it difficult to assess the associated opportunity. However, offices can lock in proper communication and take essential measures to avoid credit disruption.

There are different calculations used to estimate distributions related to cash. The introduction of computing by human beings has created a very promising way to address these types of challenges. Deep knowledge based on uncertainty and inaccuracy provides adaptive decision-making in situations where category boundaries are not clearly defined. They are different from the usual parallel basis, which deals with true or false values; the feathered inference method works widely, allowing various variables to be presented and the degree of support to be assessed. This

DOI: 10.1201/9781003538837-4

adaptability makes a good foundation, which is ideal for modeling and accounting in areas that are characterized by volatility and quality, such as support. As a part of the structure of the development support of banks, the easy-to-use basis offers a complete approach that considers several criteria at the same time and satisfies all the interrelated factors that influence our loan choices. If the primary components are combined with multi-criteria decision-making techniques, then the budget trainer will be able to make different leaps to evaluate credit requests in a more detailed and serious way. The purpose of this function is to show a complete structure that uses intelligent reasoning and multi-criteria decision-making procedures to ensure the all-around success of banks. For supporting finance, it has been a planning device for a long time, furnishing organizations and people with cash to contribute, make, and develop. People who borrow money from an individual and earn money from a business can be an important aspect of conducting financial transactions. In any event, the explanation that moving is full of risks, essentially credit issues, if we assume that the borrower delays in reimbursing the home, the bank brings around disaster, which can normally influence the abundance and dependability of the cash. Banks utilize different strategies and gadgets to reduce credit risk. One such instrument is the utilization of this fuzzy model with MCDM models to measure the probability that a borrower will default on their advancement.

4.2 LITERATURE STUDY

Fuzzy logic, along with MCDM methods, is widely used as a powerful tool for handling the inherent uncertainty and imprecision in decision-making related to finances. [1] Authors considered the significance of MCDM in optimizing portfolios, balancing conflicting objectives, and enhancing transparency in decision-making. The study presents an integrated intuitionistic Z-AHP and Z-TOPSIS methodology to address decision-makers' hesitancy and uncertainty, utilizing Z-numbers to enhance reliability [2]. Some of the methods discussed are based on intuitive fuzzy TOPSIS, intuitive fuzzy TOPSIS & DEA-AHP, intuitive VIKOR, Pythagorean WASPAS, Pythagorean ENTROPI, hesitant CBD, hesitant MABAC, triangular EDAS, triangular PROMETHEE, q-Rung Orthopair COPRAS, and fuzzy type 2 ELECTRE [3]. A fuzzy logic-based computing framework provides cooperative decision-making while avoiding inefficient use of the sensing power of smart devices [4]. A model consists of three phases: interpretative structural modeling (ISM) to construct a decision-making network; fuzzy analytical network processing (FANP) to obtain the weights of factors; and fuzzy technique for order of preference by similarity to the ideal solution (fuzzy TOPSIS) to rank the operation systems [5]. It proposes a fuzzy expert system to handle the ambiguity in the bank loan approval process. The system makes use of a fuzzy tool in MATLAB software and has two stages. The first stage uses three output parameters, that is, "repayment," "ability to manage," and "risk." The second stage uses the values of these three parameters as its input parameters to make the final decision on whether to approve the credit or not [6]. The chapter uses the Tsukamoto fuzzy method to optimize the process of assessing the feasibility of loan applications at Savings and Loan Credit with assessment criteria: loan value, total income, loan term, and collateral value [7]. The chapter uses the fuzzy logic method to calculate the creditworthiness of each prospective loaner based on the inquiries made in the field survey [8]. The chapter presents a comprehensive process considering several financial criteria and several alternatives based on several experts' (decision-makers') opinions and subjective judgments [9]. The use of machine learning algorithms on loan data to extract important information has been explained [10]. This model can be used by organizations to make the right decision about whether to approve or reject the loan request of their customers [11]. It is a comprehensive review using machine learning–based approaches for financial market prediction [12]. It analyzes efficiency of Indian life insurance companies using DEA and SEM [13]. It provides a perspective for investors and application of fuzzy-AHP in priority-based selection of financial indices.

4.3 OBJECTIVES

Here, our studies have various objectives:

I. The first objective is to develop such a decision-making structure for banks so that their loan approval systems improve. That is why we are combining fuzzy logic with multi-criteria decision-making methods in Python programming. This objective will improve the efficiency and accuracy of the loan approval process at the bank and reduce the risk of loan default, thereby increasing the financial stability of banking institutions.
II. During the loan approval process by the banks, there are many complicated and uncertain situations. So, to prevent that, the loan approval process is divided into multiple factors, each of which has its own importance. These factors include the applicant's credit score, income, employment duration, interest rate, loan term, and debt-to-income (DTI) ratio. When we understand these complexities, it will help us identify the crucial part of developing a decision-making framework that can accurately assess the risk associated with each loan application.
III. Our third objective is to include the fuzzy logic system with a multi-criteria decision-making method. Multi-criteria decision-making methods are designed to handle decisions that involve multiple, often conflicting, criteria. Combining fuzzy and MCDM methods, the bank can be able to determine the factors accurately in seconds.
IV. The fourth one is to apply Python since it is a widely used programming language that is known for its simplicity and versatility. Applying it will be a good choice for implementing the proposed framework. This application will help to show the practical application of the framework and thereby encourage banking institutions to adopt new technology.

4.4 INPUT DATASET INFORMATION

The dataset used here is a secondary data that has been taken from Coursera's Loan Default Prediction Challenge[1] for this research and model creation. It is a unique

TABLE 4.1
Membership Function

Input Variables	Terms	Range
Credit score	Poor/average/excellent	300–850
Income	Low/medium/high	0–100,000
Months employed	Short/medium/long	0–120
Interest rate	Low/medium/high	0–25
Loan term	Short/medium/long	12–60
DTI ratio	Low/medium/high	0–1

dataset, and we can test our fuzzy logic models performance using this. The dataset contains a huge number of entries but for this research we will be using only a few. Below are the factors that are generally used to assess whether an individual is eligible for a loan or not:

I. *Credit score:* It is a score given to individuals to represent their creditworthiness, that is, whether they can repay debts as per their credit history.
II. *Income:* Income suggests whether an individual is financially stable and capable of repaying debts or not.
III. *Months employed:* It is the tenure of an individual's employment, and it shows whether their income source is stable or not.
IV. *Interest rate:* It is the extra cost of borrowing a loan which depends upon the loan amount and duration of loan repayment.
V. *Loan term:* It is the duration over which the loan is repaid. Interest rate changes depending upon loan terms.
VI. *DTI ratio (debt-to-income ratio):* It is the ratio of an individual's loan to their monthly income, how much they can repay the debts from their monthly income, and it signifies their capability of repayments.

4.5 METHODOLOGY

In this section, we have combined fuzzy logic with MCDM to create a structure for bank approval system. We have created membership functions.

Fuzzification of Inputs: In this we are taking the inputs. The numerical values are given using inputs that are converted into membership function. Here these input variables are credit score, income, months employed, interest rate, loan term, and DTI ratio which are fuzzified. Here we have mentioned the triangular and trapezoidal membership functions. Using fuzzification process, we allow the system to take in this subjective information. Similarly, bank fuzzification takes place where inputs are taken which are then converted into membership functions, as provided in Table 4.1.

Fuzzy Inference Linked to Rule Base: In the next step of methodology, we develop a set of rules where fuzzy inputs are converted into fuzzy output based on the rules, as shown in Figure 4.1. These are discussed below:

(a) If the applicant's credit score is poor or their income is low, the bank will not grant them loan.
(b) Loans will be provided by the bank to the applicants if the bank finds out that their credit score is excellent and their income is high.
(c) If the bank finds out that the applicant has a short tenure of employment and a low income, then it will not grant loan.
(d) If the bank finds out that the interest rate and debt-to-income ratio are high, it will not approve the loan.
(e) If the bank finds out that the loan term is short and the applicant's income is low, it will not grant the loan.

Defuzzification Process: Under this process, input variables are analyzed, and final outputs are given. According to combined final outcomes, the final decision is taken which is precise, that is, whether to give him/her loan or not.

Implementation and Result Analysis: The final step in the methodology is the implementation of the fuzzy logic system and the analysis of the results. The document mentions that the system is implemented in Python, a widely used programming language known for its simplicity and versatility. The performance of the system is tested using real-world data. The range of approval percentages varies significantly, highlighting the diverse nature of the applicant pool, which consists of individuals with varying levels of creditworthiness. This step is crucial as it validates the effectiveness of the fuzzy logic system in making accurate and efficient loan approval decisions. This methodology includes very organized research work. We have worked very meticulously to develop the decision-making framework for the banking system of loan validation. We have combined fuzzy logic, MCDM, and a Python programming platform to develop a flexible solution for the bank. This process will help banks to reduce defaults. Here we have used Python because Python is a very simple and versatile programming logic. It is easier to use than other programming languages as we can solve complex parts of each model. Here we have used real data to show the practicality of the model. The findings of the study will provide valuable insights for both academics and practitioners in the field of banking.

4.5.1 Algorithm

STEP 1: FUZZIFICATION OF INPUTS

```
def fuzzify_inputs(inputs):
# Transform the input numerical values into membership functions

# The input variables include credit score, income, months
employed, interest rate, loan term, and DTI ratio

# Use triangular and trapezoidal membership functions for
input fuzzification
pass
```

STEP 2: FUZZY INFERENCE LINKED TO RULE BASE

```
def fuzzy_inference(rule_base, fuzzified_inputs):

# Develop a set of fuzzy rules that capture the complex relationships between the input criteria and the loan approval outcomes

# Translate the fuzzy input variables into a fuzzy output that represents the loan approval decision
pass
```

STEP 3: DEFUZZIFICATION PROCESS

```
def defuzzify(fuzzy_output):

# Transform the aggregated fuzzy output into a precise value

# The defuzzification process ultimately determines the loan approval decision with clarity and precision based on the fuzzy logic computations
Pass
```

STEP 4: IMPLEMENTATION AND RESULT ANALYSIS

```
def implement_and_analyze(fuzzified_inputs, rule_base):

# Implement the fuzzy logic system and analyze the results

# Test the performance of the system using real-world data
pass
```

4.5.2 Pseudocode

```
# Main function
def main():

# Define the inputs and the rule base
inputs = {. . .}
rule_base = {. . .}

# Fuzzify the inputs
fuzzified_inputs = fuzzify_inputs(inputs)

# Perform the fuzzy inference
fuzzy_output = fuzzy_inference(rule_base, fuzzified_inputs)

# Defuzzify the output
precise_output = defuzzify(fuzzy_output)

# Implement the system and analyze the results
implement_and_analyze(fuzzified_inputs, rule_base)
return precise_output

# Call the main function
main()
```

4.5.3 Flowchart

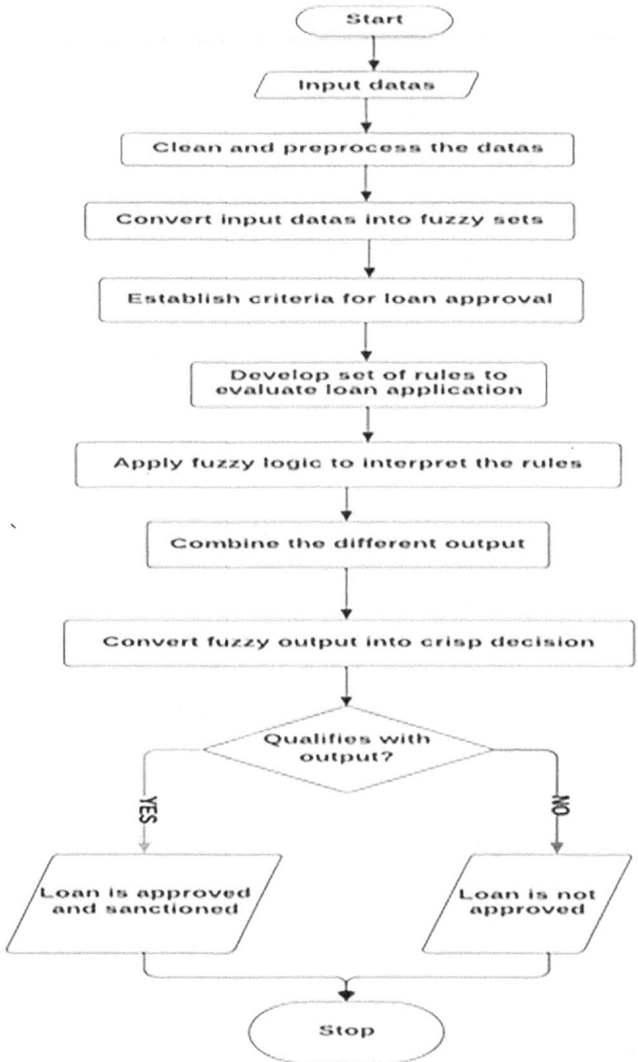

FIGURE 4.1 Flowchart of the proposed model.

4.6 RESULT ANALYSIS

To investigate alternative models that can assist loan officers in retail credit evaluation, we implemented a credit evaluation system based on fuzzy logic using bank data. This chapter presents the results of applying the fuzzy logic system to a set of loan applications. Each application is assigned a loan approval score, a numerical value that represents whether it will get approved or not. The scores range from 16 to 60, indicating a wide variety of creditworthiness of the applicants as assessed by our model in Python. Here are the loan approval scores for each application:

Loan ID: I38PQUQS96 | Loan Approval: 22.592592592592588
Loan ID: HPSK72WA7R | Loan Approval: 18.46367521367521
Loan ID: C1OZ6DPJ8Y | Loan Approval: 18.104477611940297
Loan ID: V2KKSFM3UN | Loan Approval: 20.271769962896382
Loan ID: EY08JDHTZP | Loan Approval: 16.674453564613195
Loan ID: A9S62RQ7US | Loan Approval: 60.83301406033403
Loan ID: H8GXPAOS71 | Loan Approval: 17.188702669149595
Loan ID: 0HGZQKJ36W | Loan Approval: 23.452352708580463
Loan ID: CM9L1GTT2P | Loan Approval: 19.75845410628019
Loan ID: IA35XVH6ZO | Loan Approval: 50.00124998958311
Loan ID: Y8UETC3LSG | Loan Approval: 17.499999999999996
Loan ID: RM6QSRHIYP | Loan Approval: 17.048507235475316
Loan ID: GX5YQOGROM | Loan Approval: 16.706349206349202
Loan ID: O5DM5MPPNA | Loan Approval: 17.291819291819287
Loan ID: ZDDRGVTEXS | Loan Approval: 23.41265597147951
Loan ID: 9V0FJW7QPB | Loan Approval: 20.942254152153065
Loan ID: O1IKKLC69B | Loan Approval: 22.70810313075505

The loan approval scores can be interpreted as a measure of the likelihood of loan approval, with higher scores indicating a higher likelihood of approval. For instance, the loan application with ID "A9S62RQ7US" has the highest approval score of approximately 60.83, suggesting that this applicant is highly likely to have their loan approved. On the other hand, the loan application with ID "EY08JDHTZP" has the lowest approval score of approximately 16.67, suggesting that this applicant is less likely to have their loan approved. The range of approval rates varies significantly from 16.67% to 60.83%, showing a wide range which in turn tells that applicants have varying creditworthiness which is practical in terms of real world. Most of the approval rates lean toward lower values while some show higher rates, so we can say the distribution is skewed. Approval rates are related to several factors like credit score, income, length of employment, debt-to-income ratio, etc. Individuals with higher credit scores, stable employment records, and lower levels of debt are more likely to obtain approval for a loan.

The horizontal axis indicates the credit score, which ranges from 300 to 850. On the vertical axis, the level of membership in fuzzy sets is displayed, ranging from 0.0 to 1.0. The *y*-axis is labeled with terms such as "poor," "average," and "excellent, as shown in Figure 4.2.

Poor (0.0–0.6): This fuzzy set includes credit scores ranging from 300 to approximately 580. It probably signifies a minimal likelihood of loan approval. *Average (0.2–0.8):* This fuzzy set encompasses credit scores from about 580 to 740. It probably signifies an ambiguous loan approval status, with other factors being considered. *Excellent (0.4–1.0):* This fuzzy set consists of credit scores from around 740 to 850. It probably signifies a strong likelihood of loan approval.

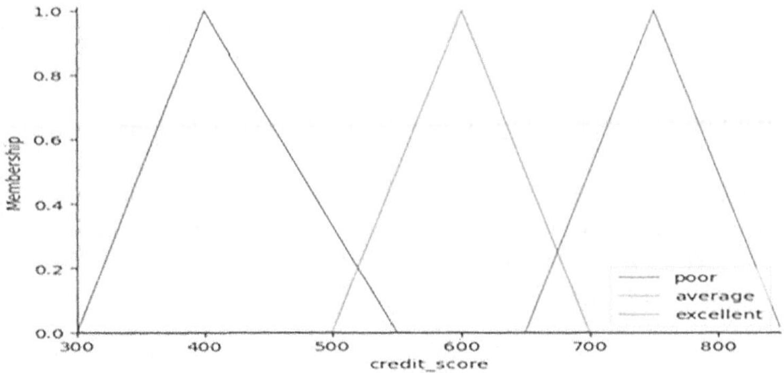

FIGURE 4.2 Credit score versus membership of card plot as per poor average and excellent criteria.

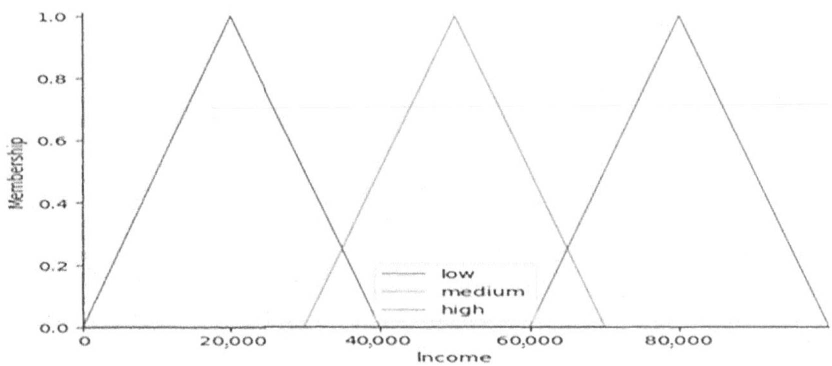

FIGURE 4.3 Income versus membership of card plot as per low, medium, and high range.

The number of individuals is represented on the *y*-axis, whereas income is represented on the *x*-axis. The income ranges from 0 to 80,000, as shown in Figure 4.3.

The medium-income level has a larger population than both the low- and high-income levels, indicating a potential imbalance in income distribution within the dataset.

It appears that there is a greater proportion of employees who have been with the company for a shorter period (between 0 and 20 months). This observation may suggest a significant turnover rate during the early stages of employment. As the duration of employment progresses, the number of employees seems to decline gradually, as shown in Figure 4.4.

Low (0.0–0.6): This fuzzy set denotes loan amounts ranging from 0 to approximately 10, suggesting a comparatively lower average interest rate. *Medium (0.2–0.8):* Encompassing loan amounts between approximately 10 and 15, this fuzzy set likely signifies an average interest rate. *High (0.4–1.0):* Covering loan amounts

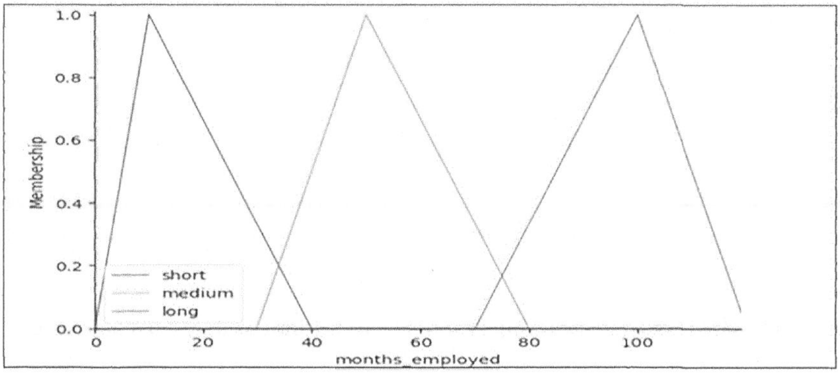

FIGURE 4.4 Months employed versus membership of card plot as per short, medium, and high range.

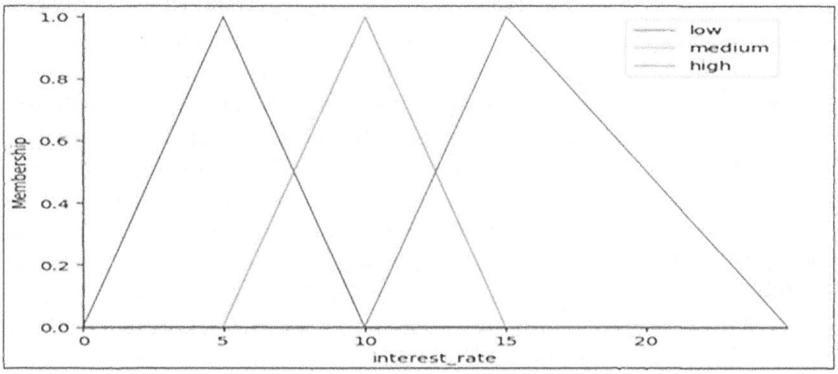

FIGURE 4.5 Interest rate versus membership of card plot as per low, medium, and high range.

from around 15 to 20, this fuzzy set indicates a higher average interest rate, as shown in Figure 4.5.

A concentration of data points is evident in the lower left corner of the graph, where loan terms are shorter, and membership falls under the "short" category. This implies that a significant portion of loans have shorter terms and belong to the "short" category. Additionally, scattered data points across the rest of the graph suggest the presence of loans with diverse terms and membership categories. It appears that the mean line is mostly flat, while the average line has a peak in the middle around a dti_ratio of 0.5. The provided output suggests as per Figure 4.6 that the proposed methodology can effectively apply fuzzy logic and multi-criteria decision-making methods to the loan approval process, resulting in a nuanced assessment of the creditworthiness of applicants. However, further analysis would be needed to fully evaluate the performance of the system and its potential to enhance the loan approval process.

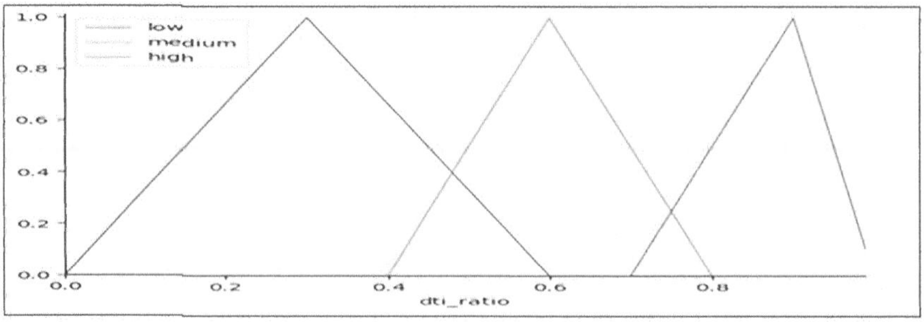

FIGURE 4.6 DTI versus membership of card plot as per low, medium, and high range.

4.7 LIMITATIONS

The selection of fuzzification's membership functions is another important constraint. Although trapezoidal and triangle membership functions are used in the document, other forms may have been more suitable based on the particulars of the input variables. It can take some time and be difficult to define these functions, especially if there are a lot of input variables. The decision-making process can also be greatly impacted by the defuzzification procedure, which is essential to turning the fuzzy output into an exact number. The choice of defuzzification method can yield different results, and the most appropriate method may depend on the specific requirements of the system. The performance of the system, tested using real-world data, can be significantly impacted by the quality and availability of this data. If the data is incomplete, outdated, or biased, it could lead to inaccurate or misleading results. Furthermore, the system's performance may be affected by the specific characteristics of the data used for testing. These limitations highlight areas that could be addressed in future work, and they provide valuable insights for the ongoing development of more effective decision-making tools and strategies in the field of banking. These limitations do not necessarily undermine the value of the research, but they do highlight areas that could be addressed in future work. It is also important to note that acknowledging and understanding these limitations can help guide the practical application of the system and inform future research in this area.

4.8 CONCLUSIONS

This study provides a thorough plan that combines fuzzy logic with multi-criteria decision-making techniques to speed up bank loan approval processes. The main aim of this work is to improve the loan approval process's accuracy and efficiency, which will reduce default risk and strengthen banking institutions' financial stability. The process of approving a loan involves several variables, each with a different level of relevance and volatility. To manage these complexities and uncertainties, a system that can handle them must be developed. Fuzzification of inputs, fuzzy inference related to the rule base, defuzzification process, implementation, and outcome analysis are some of the essential components of research techniques. The fuzzification of inputs is the initial step in the computation of fuzzy logic, transforming the numerical values of the input into membership functions. The fuzzy inference process, linked to the rule base, develops a set of fuzzy rules that capture the complex relationships between the input criteria and the loan approval outcomes. The fuzzy output is converted by using the defuzzification process into an understandable value, which allows the fuzzy logic computations to be used clearly and precisely to determine the decision of loan acceptance. The outcomes demonstrate the wide range of people in the applicant pool, each with a different creditworthiness score. This explains the framework's usefulness and shows how it can improve the precision and effectiveness of the loan approval process. However, the research also acknowledges potential limitations such as subjectivity in rule creation, choice of membership functions, method of defuzzification, data quality and availability, scalability and performance, generalizability, and validation. These limitations provide valuable insights for the ongoing development of more effective decision-making tools and strategies in the field of banking. They highlight areas that could be addressed in future work and provide valuable insights for the ongoing development of more effective decision-making tools and strategies in the field of banking.

NOTE

1 *Source:* https://www.coursera.org/projects/data-science-coding-challenge-loan-default-prediction and download: https://www.kaggle.com/datasets/nikhil1e9/loan-default/data

REFERENCES

1. Kahraman, Cengiz, and Elif Haktanır. "Fuzzy Multi-Criteria Investment Decision Making." In *Fuzzy Investment Decision Making with Examples*, pp. 223–244. Cham: Springer Nature Switzerland, 2024.
2. Kahraman, Cengiz. "Multi-Criteria Decision Making Methods and Fuzzy Sets." In *Fuzzy Multi-Criteria Decision Making: Theory and Applications with Recent Developments*, Vol. 16, pp. 1–18. Springer Science & Business Media, 2008.
3. Haseeb, Khalid, Irshad Ahmad, Mohammad Siraj, Naveed Abbas, and Gwanggil Jeon. "Multi-Criteria Decision-Making Framework with Fuzzy Queries for Multimedia Data Fusion." *ACM Transactions on Asian and Low-Resource Language Information Processing* 86 (2024): 1413–1423.
4. Lee, Amy H. I., and He-Yau Kang. "A Three-Phased Fuzzy Logic Multi-Criteria Decision-Making Model for Evaluating Operation Systems for Smart TVs." *Applied Sciences* 13, no. 13 (2023): 7869.
5. Setiawan, Tabah Heri, and Laras Prihatini. "Tsukamoto Fuzzy in Optimizing the Creditworthiness Assessment Process at Savings and Loan Cooperatives." *Barekeng: Jurnal Ilmu Matematika dan Terapan* 17, no. 2 (2023): 0775–0786.
6. Ginting, S. L. B., M. R. M. Rizky, and Y. R. Ginting. "The Application of Fuzzy Logic Method in the Debtors Eligibility Assessment System of Microfinance Institution." In *IOP Conference Series: Materials Science and Engineering*, vol. 879, no. 1, p. 012039. IOP Publishing, 2020.
7. Aphale, Amruta S., and Sandeep R. Shinde. "Predict Loan Approval in Banking System Machine Learning Approach for Cooperative Banks Loan Approval." *International Journal of Engineering Trends and Applications (IJETA)* 9, no. 8 (2020).
8. Chaudhuri, Tamal Datta, and Indranil Ghosh. "A Multi-Criteria Decision Making Model-Based Approach for Evaluation of the Performance of Commercial Banks in India." *IUP Journal of Bank Management* 13, no. 3 (2014): 23.

9. Santra, Soumen, and Rohit Mondal. "A New Sorting Approach for Big Data Set: Hit Sort." In *2017 International Conference on Information, Communication, Instrumentation and Control (ICICIC)*, pp. 1–5. IEEE, 2017.
10. Tarafder, Mousumi, Soumen Santra, Sweta Sharma, and Arpan Deyasi. "Evaluation of Optimum Batting Order Based on Partnership Analysis Using Ant Colony Optimization Technique." In *International Conference on Advances in Data Science and Computing Technologies*, pp. 227–233. Singapore: Springer Nature Singapore, 2022.
11. Nandi, Bhaskar, Subrata Jana, and Krishna Pada Das. "Machine Learning-Based Approaches for Financial Market Prediction: A Comprehensive Review." *Journal of AppliedMath* 1, no. 2 (2023): 134–134.
12. Ghosh, Arijit. "Analyzing Efficiency of Indian Life Insurance Companies Using DEA and SEM." *Turkish Journal of Computer and Mathematics Education (TURCOMAT)* 12, no. 12 (2021): 3897–3919.
13. Jana, Subrata, Bibhas Chandra Giri, Anirban Sarkar, Chiranjibe Jana, Željko Stević, and Marko Radovanović. "Application of Fuzzy AHP in Priority Based Selection of Financial Indices: A Perspective for Investors." *ECONOMICS-Innovative and Economics Research Journal* 12, no. 1 (2024): 1–27.

5 Policy Implications and Regulatory Framework

Anshit Mukherjee, Biswadip Basu Mallik, Gunjan Mukherjee, and Sudeshna Das

5.1 INTRODUCTION

Circular economy (CE) (Allen and Garmestani 2015) is a revolutionary paradigm whose ambition is to address the issues created due to global sustainability hurdles in particular like change in climate, depletion in resources, and pollution, to name but a few. Actually, circular economy can be stated as paradigms and regulations that continuously bolster the proper utilization of resources through actions like reusing, repairing, refurbishing, and recycling. This approach completely differs from traditional linear economy (Bansal and Roth 2000) and this traditional linear economy can be defined as "take–make–disposal" architecture, which produces noticeable waste and thus degrading our environment. Circular policies are important because of their capability to decouple growth of economy from resource consumption and thus promoting sustainable growth and improving resilience across different domains.

Historically, the transformation of CE policies can be routed to the early 2000s, gaining pace with the recognition of sustainability as a critical global agenda. The European Union has been at the forefront, implementing comprehensive CE strategies, while countries like China adopted top-down strategy to assess CE principles onto national policy frameworks. In the literature review section, we found that many nations across the globe have taken steps to initiate various circular economy policies, but a void is still existing between the implementation and effectiveness and the scenario further worsens due to regulatory uncertainties and lack of proper stakeholder collaboration (Bocken et al. 2014). The serious desire to improve the circular economy policies has highly motivated us to choose this topic for further exploration to provide a more sustainable economy technique. The detected void in existing literature includes regulatory fragmentation, insufficient stakeholder engagement, and the complexities of measuring the effectiveness of CE initiatives. The main aim of this work is to address these voids by analyzing current CE policies, exploring successful case studies, and proposing recommendations for enhancing regulatory frameworks. The objectives of this chapter are to state and contextualize CE policies within the broader sustainability discourse, to examine the historical evolution and current state of CE policies worldwide, to identify and analyze the voids faces in the implementation of CE policies, and to propose actionable strategies for improving stakeholder collaboration and regulatory coherence. Figure 5.1 shows a pie chart that divulge percentage of contribution of different countries worldwide in implementing

DOI: 10.1201/9781003538837-5

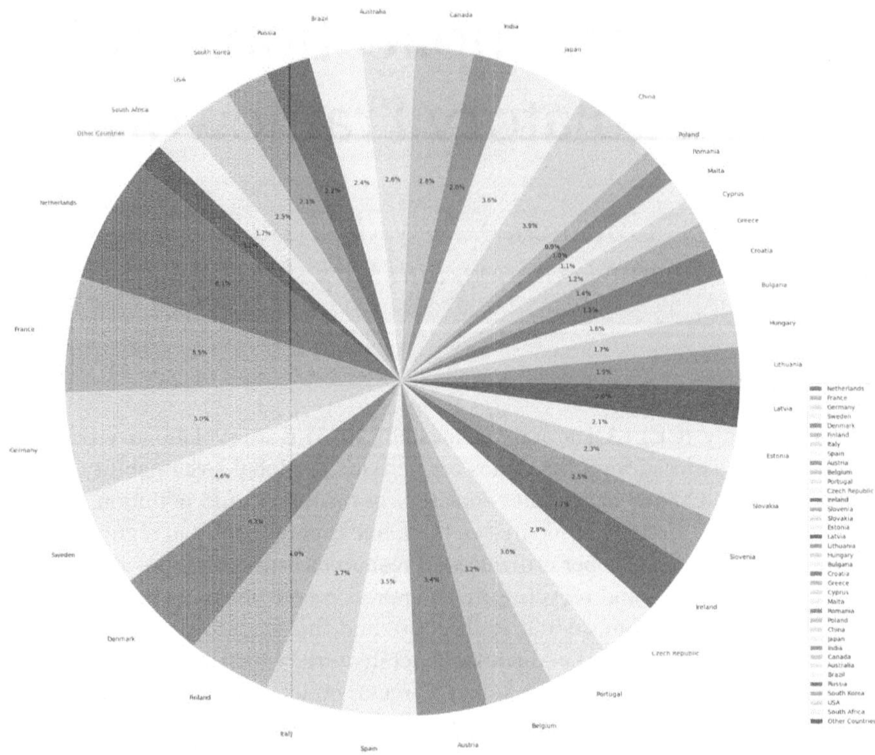

FIGURE 5.1 A pie chart showing proportion of countries implementing circular economic policies globally as of 2023.

various circular economy policies globally with respect to the year 2023 (Circle Economy 2023).

5.2 LITERATURE REVIEW

We have been able to find out noticeable gap in the empirical testing of different circular economy paradigms, in particular in the context of dynamic industries. Also, there is absence of attention on the social and cultural aspects of circular economy which is crucial to mastery the proper relation between behavior of consumer and stakeholder engagement. We have been deeply motivated by this gap to pursue research that not only look after these aspects but also at the same time help in constructing a more robust paradigm that includes both these dimensions of circular economy. This chapter divulge to address these gaps which helps in proper understanding of circular economy. J.A.M. Albrecht (Albrecht et al. 2024) in 2024 made a qualitative examination and their work highlighted the need of incorporating social dimensions in circular economy practices. The disadvantage of their work is that the limited samples they have may not represent broader trends. Kapoor Nair et al. (2024)

in 2024 have performed different examination of different case studies to explain the successful implementation of circular economy in different sectors. The disadvantage of their work is that their results are not generalizable with respect to different contexts. Sinha et al. (2024) in 2024 have conducted meta-analysis which computes the effect of circular economy principles across various metrics of sustainability. The researches warn that the potential biases in selected studies can impact their findings. Mehta et al. (2024) in 2024 performed comparative examination to distinguish circular economy principles across different countries in this year and their findings indicate that the variability in information sources often lead to several inconsistencies. Joshi et al. (2024) in 2024 performed qualitative interviews to identify different perspectives of stakeholder on circular economy policies and surprisingly their findings are greatly influenced by subjective interpretations. Halder et al. (2024) in 2024 made an in-depth study to calculate the effect of circular economic principles with the passage of time and one thing to note is that this technique is time-consuming and also depend on intensive resources. J.H.M. and his team performed a comparative literature review in 2024 where they tried to create a distinguished computation of circular economy practices in urban versus rural settings. The disadvantage of their work is that they did not focus in depth on specific case studies. Chirag et al. (2024) in 2024 have made a systematic review to find out regulatory paradigm that can support the initiatives of circular economy and this study actually does not have any practical recommendations for policymakers.

Geng (2023) and his team in 2023 put forward a comprehensive paradigm that can make a conceptual investigation on circular economy principles. The disadvantage of their work is that the paradigm is actually too broad to tackle any particular application. Lee et al. (2023) in 2023 mentioned interdisciplinary approaches on circular economy in their review article. The disadvantage of their work is that they overlooked the localized practices and policies. Gupta et al. (2023) in 2023 adopted a mixed technique to compute the potentiality of circular economy principles in India. This approach of focusing on particular area limits their applicability to any other domain. Oliveira and his team (2023) in 2023 performed a systematic review and this paper identified the voids in circular economy research and gave some future guidelines. The disadvantage of this paper is that it lacks actionable guidelines for the practitioners. Chiu et al. (2023) in 2023 performed an empirical examination and they identified the behavior of consumer toward the circular products. The disadvantage of their work is that their size of sample does not tackle broader demographics. Wang et al. (2023) in 2023 put forward before us successful circular habits in the domain of manufacturing through a case study. The disadvantage of their work is that the context-specific results of their work may not be always applicable universally. Ferreira et al. (2022) in 2022 actually performed literature review to identify the main public policies and the appropriate analytical weapons in circular economy. The disadvantage of their work is that their examination was limited to the period 2017–2021, thereby missing the recent strategies. Singh et al. (2022) in 2022 indicated the potential challenges and opportunities in circular economy practices through a proper literature review, mainly pinpointing on emerging trends. The disadvantage of their work is that their work lacks empirical information. Khan et al. (2022) in 2022 also talked about the hurdles and scopes of implementing circular economy on developed countries. The disadvantage of their work is that this approach only focusing on developed countries may not accurately pinpoint the global trends.

5.3 REGULATORY FRAMEWORKS SUPPORTING CIRCULAR ECONOMY

There are various regulatory paradigms which boost circular economy from region to region, thus pinpointing the heterogeneous approach for proper handling of resources, diminishing of wastes, and sustainable habits. Now let us dive deep into the noticeable regulatory frameworks in prominent regions like European Union, China, the United States, Japan, India, and others that give boost to circular economy in the present arena.

One of the remarkable stone of the European Green Deal is the Circular Economy Action Plan (CEAP) (European Commission 2020a) which was adopted in March 2020. The main emphasis is to promote sustainable production and consumption by incrementing resource efficacy and reducing waste. The CEAP (World Business Council for Sustainable Development 2020) has set a number of measures that cover the entire life cycle of products, from design to disposal. The key initiatives include eco-design requirements (the plan made modifications on existing regulations to ensure that products are designed for durability, reparability, and recyclability), packaging waste regulations (it aims to reduce packaging waste through stricter regulations on single-use plastics and increased recycling targets), and market for secondary raw materials (the plan that seeks to create a well-functioning market for secondary raw items which also includes the building of EU-wide end-of-waste criteria). The CEAP stresses cooperation among public authorities, businesses, and civil society to fulfil its goals. It also aspires to decouple economic growth from resource use, promoting a regenerative growth model that reduces waste and environmental impact. This plan is instrumental in setting a global standard for circular economy practices and made a significant role in the transition of European Union to a sustainable economy. The European Green Deal (European Commission 2020b) is an all-inclusive policy model whose aim is to make Europe climate neutral by 2050 (Chatham House 2024). It includes circular economy principles into broader environmental goals which includes biodiversity protection and pollution reduction. The Green Deal helped European Union in adopting different initiatives in boosting the circular economy. These initiatives are investment in green technology, regulatory approaches, and public engagement. This deal has taken the role of a guiding paradigm for the different environmental policies of EU.

The 14th Five-Year Plan for circular economy (Wang and Zhang 2020) by China focuses on incrementing resource utilization and promotion of recycling and green product design. The noticeable aspects of this plan include strategic targets (the plan gives ambitious targets such as resource productivity and waste reduction by 2025, emphasizing to increase recycling rate of key materials) and investment in green technologies (encouraging the development of green technologies to support circular practices across industries). This plan reflects China's commitment in transitioning circular economy as part of its broader environmental and economic strategies. China's Circular Economy Law sets a legal framework for promotion of circular practices which focuses on waste management (improving waste management systems to boost recycling and resource recovery) and incentives for circular practices (offering incentives for businesses adopting circular practices which includes tax breaks and subsidies). This law is playing an important role for the implementation of circular economy principles at the national level of the country.

The Bipartisan Infrastructure Law (U.S. Congress 2021) of the United States provides significant funding for improving recycling infrastructure and accessibility. The noticeable components of this law include investment in recycling facilities (funding for the construction and upgradation of recycling facilities to improve capacity) and accessibility improvements (initiatives whose motive is to make recycling programs more accessible to communicate across the country). This law holds a decade ahead step in improving waste management and resource recovery in the United States. The Recycling Infrastructure and Accessibility Act (U.S. Environmental Protection Agency 2021) aims to bolster recycling programs through improving efficacy (enhancing the overall efficacy of waste management systems through better technology and practices) and public awareness campaigns (promoting awareness and education for recycling and waste reduction). These legislative efforts are crucial in advancing circular economy practices in the United States.

The Basic Act on Establishing a Circular Society (Japan Ministry of the Environment 2020) empowers the reuse and recycling of assets in Japan. The key goals of this act incorporate diminishing natural effect (pointing to minimize environmental impact through sustainable practices) and empowering public participation (engaging citizens in recycling and resource conservation efforts). This enactment is a foundational element of Japan's strategy to promote circular economy practices.

Australia has set up National Waste Policy (Australian Government 2020) to achieve the ambitious targets for recovery of wastes and recycling whose motive is to recover 80% of the total waste by 2030. This noticeable step includes waste reduction strategies (implementation of new methodologies to minimize waste generation across sectors) and investment in recycling technologies (supporting the development of advanced recycling technologies). The Product Stewardship Laws are also encouraging manufacturers in taking initiatives for entire life cycle of their products and the noticeable feature includes extended producer responsibility (mandating that producers manage the disposal and recycling of their products) and incentives for sustainable practices (providing incentives for businesses that adopt circular practices).

The National Resource Efficiency policy (Government of India 2021) of India provides significant attention for improving resource efficiency and minimizing waste generation. The noticeable components of this law include sustainable practices (encouragement of sustainable practices across various sectors, including manufacturing and agriculture) and capacity-building (encouraging the capacity of stakeholders to implement circular practices). This strategy is playing an important role in transmitting India to a circular economy.

Brazil has set up National Circular Economy Policy (Brazilian Government 2020) to encourage the sustainable production strategies and reduce environmental impact.

TABLE 5.1
Comparison of Administrative Approaches over Jurisdictions

Region/Country	Key Policies	Objectives
European Union	Circular Economy Action Plan, European Green Deal	Advance sustainable production and utilization; move forward resource efficiency
China	14th Five-Year Plan for Circular Economy, Circular Economy Law	Maximize asset utilization; advance reusing and waste reduction
The United States	Bipartisan Infrastructure Law, Recycling Infrastructure Act	Enhance reusing and decrease waste; construct infrastructure
Japan	Basic Act on Establishing a Circular Society	Support material reuse and recycling; reduce environmental impact
Australia	National Waste Policy, Product Stewardship Laws	Achieve 80% waste recovery by 2030; encourage recycling
Canada	Zero Plastic Waste Strategy	Eliminate plastic waste; promote alternative materials
India	National Resource Efficiency Policy	Make strides resource effectiveness; diminish waste generation
Brazil	National Circular Economy Policy	Promote sustainable production methods; reduce environmental impact
South Korea	Framework Act on Resource Circulation	Advance resource circulation; minimize waste
The United Kingdom	Resource and Waste Strategy	Improve recycling rates; reduce landfill

This noticeable step includes regulatory framework (establishing regulations that provide circular practices in various industries) and public awareness campaigns (promoting awareness of circular economy principles among businesses and consumers). This model is critical for improving circular economy initiatives in Brazil. Table 5.1 gives a comparative table of administrative approaches over different jurisdictions for encouragement of circular economy.

5.4 POLICY INSTRUMENTS FOR CIRCULAR ECONOMY IMPLEMENTATION

The shift to circular economy requires a mixture of policy instruments that generate the right incentives and authorize the necessary modifications in productions and consumption patterns (Stahel 2016). These policy instruments can be broadly divided into three main types: economic instruments, informational instruments, and voluntary agreements.

5.4.1 Economic Instruments

Economic instruments (Tanskanen and Kallio 2018), like taxes, subsidies, and incentives, are influential weapons for encouraging circular economy practices. The mechanism of this strategy is to alter the relative prices of products and services, making circular options more attractive to businesses and consumers. Some critical exemplars of economic instruments are as follows:

- *Taxes on virgin materials and landfill squander:* If the costs of utilizing virgin materials and arranging of squander in landfills is raised at that point, this step will energize the utilization of reused materials and appropriation of circular commerce models.
- *Subsidies for circular exercises:* We ought to energize exercises like item repair, remanufacturing, and reusing by giving monetary help and this step will offer assistance in overcoming the initial cost barriers and stimulate investment in circular solutions.
- *Extended producer responsibility (EPR) schemes:* It ought to be the obligation of the producers to take responsibility for end-of-life management of their products which incorporates monetary commitments, incentivizes in eco-design, and the improvement of take-back frameworks.

5.4.2 Informational Instruments

The objective of informational instruments (van Buren et al. 2016) like awareness campaigns and educational programs is to change attitudes and behaviors by providing information and knowledge about the benefits of circular economy practices. Some noticeable exemplars of informational instruments are as follows:

- *Public awareness campaigns:* If we uphold the advantages of circular economy before the common public, this will increase demand for sustainable goods and services and encourage behavioral changes.
- *Eco-labeling and product information:* If we provide clear and reliable information about the environment, the performance and circularity of products can help consumers in making informed decisions and drive demand for circular solutions.
- *Circular economy education:* If we amalgamate circular economy principles into school curriculum, then through professional training, we can develop the required skills and knowledge for the transition of circular economy.

5.4.3 Voluntary Agreements

Voluntary agreements (Wastling and Dwyer 2018) and associations between partners can complement financial and instructive instruments by cultivating collaboration and shared obligation for the move to a circular economy. Some noticeable exemplars of voluntary agreements are as follows:

- *Industry-led activities:* Businesses can come together to create voluntary guidelines, code of conduct, and best hones that advance circular economy hones inside their sectors.
- *Public–private partnerships:* Intermingling between government, industry, and civil society offer assistance in adjusting interests, share risks and benefits, and create novel arrangement for the circular economy.

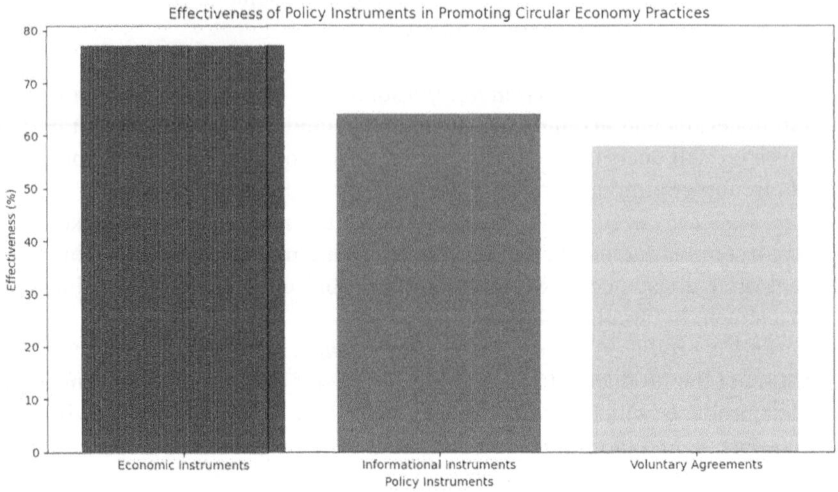

FIGURE 5.2 A bar chart showing efficacy of policy instruments in encouraging circular economy practices as per survey results conducted by EEA 2023.

- *Stakeholder platforms and networks:* Bringing together different partners, such as businesses, analysts, policymakers, and NGOs, can encourage information sharing, collaboration, and the improvement of a shared vision for the circular economy.

The efficiency of these policy instruments in encouraging circular economy practices can change depending on the particular setting and the combination of instruments utilized. A well-designed arrangement blend that combines financial, instructive, and voluntary instruments can make deliberate steady environment for the move of the circular economy. The effectiveness of the above-stated policies can understand the results of a survey which was conducted by the European Environment Agency (EEA) in the year 2023 (European Environment Agency 2023a) which questioned the expertise to rate the effectiveness of different policy instruments on the scale of 1–100 and the results are presented in Figure 5.2.

5.5 ROLE OF GOVERNMENT AND POLICYMAKERS

Governments play a vital part in encouraging the move to a circular economy through different obligations:

- *Policy advancement:* Governments are dependable for making and actualizing arrangements that empower circular hones. This incorporates setting up administrative systems, setting targets for squander decrease, and advancing feasible generation and consumption.
- *Financial bolster:* Giving monetary motivations, such as awards, appropriations, and assess breaks, can energize businesses to encourage circular

practices. Governments can moreover contribute in framework that underpins reusing and squander management.

- *Research and development:* Governments can finance and inquire about advancement activities that center on circular economy advances and practices. This incorporates supporting advancement in materials science, reusing innovations, and maintainable design.
- *Public mindfulness and instruction:* Raising open mindfulness benefits the circular economy. Governments can actualize instructive programs to educate citizens about feasible practices and energize cooperation in reusing and squander decrease initiatives.
- *Collaboration and organizations:* Governments can cultivate collaboration between different partners, counting businesses, non-governmental organizations (NGOs), and the scholarly world, to make a shared vision for the circular economy and encourage information sharing.
- *Monitoring and assessment:* Building up measurements to a degree that the adequacy of circular economy arrangements is significant. Governments ought to screen advance and assess the effect of their activities to guarantee nonstop enhancement.

Some exemplars of successful government-led initiatives are as follows:

> It is the dream and commitment of the New York city of the United States (New York City Government 2015) to accomplish zero waste by 2030. Under Mayor Bill de Blasio, NYC has taken stern initiatives and new steps in encouraging recycling and minimizing waste, which includes compulsive recycling laws and extended product responsibility (EPR) for wastes generated from electronic goods. The city put forward efforts in organization of campaigns to aware people of circular habits.

The capital city of Netherlands, Amsterdam (City of Amsterdam 2020), has also encouraged sustainable economic growth and minimization of wastes obeying a circular economy action habit. Amsterdam has put efforts in maximizing recycling rates, and increasing cooperation between public and private players. The main noticeable initiative of this program includes food waste reduction and association of local businessman to boost circular habits.

Ellen MacArthur Foundation (Ellen MacArthur Foundation 2021) in the United Kingdom has taken initiatives to guide businesses and designers in making products that can be reused, refurbished, and recycled. This guideline is popularly called as Circular Design Guide. The most remarkable achievement of this initiative is its adoption by various companies like Google and Unilever to improve design procedure of products and minimize their impact on environment.

The Finnish government (Sitra 2020) has adopted remarkable strategies and set a roadmap to create a circular economy by 2025. The government has taken remarkable actions in encouraging circular habits across various sectors which include waste management, sustainable production, and resource efficiency. This roadmap stresses cooperation between the government, businesses, and research institutions to drive innovation and implementation.

Scotland (Zero Waste Scotland 2023) has also stressed on circular economy approach for minimizing waste, incrementing recycling rates, and promotion of sustainable consumption. The government has taken initiatives to start Deposit Return Scheme for beverage containers which gives incentives for recycling and promotes minimization of littering.

The Zero Waste Masterplan of Singapore (Singapore Ministry of the Environment and Water Resources 2019) is an exemplar that is more effective in diminishing wastes sent to landfills and encouraging resource efficacy. The strategy includes efforts in expansion of recycling programs, promotion of sustainable practices, and building circular economy framework that promotes innovation in successful management of wastes.

In spite of the advance made, policymakers confront a few challenges in actualizing circular economy policies:

- *Financial imperatives:* Restricted monetary assets can prevent the improvement and execution of circular economy activities. Governments may battle to designate adequate subsidizing for fundamental foundation and bolster programs.
- *Regulatory complexity:* The existing administrative scene can be complex and divided, making it troublesome for businesses to explore and comply with circular economy policies.
- *Stakeholder engagement:* Locks in differing partners, counting businesses, shoppers, and neighborhood communities, can be challenging. Policymakers must guarantee that all voices are listened and that arrangements are inclusive.
- *Data accessibility:* A need of solid information on asset streams, squander era, and reusing rates can hinder the improvement of compelling approaches. Policymakers require exact information to make educated decisions.
- *Behavioral alter:* Empowering shoppers and businesses to embrace circular hones requires critical behavioral alter. Policymakers must actualize successful mindfulness campaigns and instructive programs to cultivate this change.
- *Short-term center:* Numerous businesses prioritize short-term benefits over long-term supportability, making it troublesome to execute circular economy hones that may require forthright ventures.

Figure 5.3 shows the role of various government levels in implementing CE policies.

5.6 STAKEHOLDER ENGAGEMENT AND COLLABORATION

Multi-stakeholder collaboration (UNEP 2021) is fundamental for the fruitful move to a circular economy. This collaboration includes different on-screen characters, including government entities, industry agents, the scholarly world, and respectful society. Each partner brings one-of-a-kind viewpoints, assets, and ability that are crucial for tending to the complex challenges related with transitioning from a linear to a circular economy.

- *Government:* Governments (World Economic Forum 2021) play a basic part in setting administrative systems, giving budgetary motivating

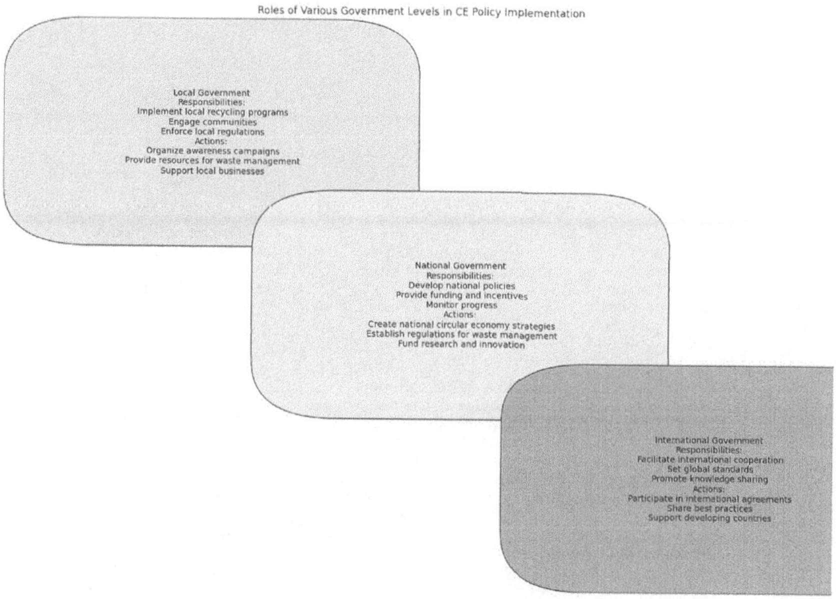

FIGURE 5.3 Flowchart displaying roles of government at different stages in implementation of circular economy policy habits.

forces, and encouraging open mindfulness campaigns. Their inclusion guarantees that approaches are adjusted with the broader objectives of supportability and circularity. Governments can, moreover, act as go-between, bringing together distinctive partners to cultivate discourse and collaboration.

- *Industry:* Businesses (European Commission 2021) are at the cutting-edge of actualizing circular hones. They have the capacity to enhance and overhaul items and forms to minimize squander and maximize asset productivity. Industry collaboration can lead to the advancement of shared measures, best practices, and joint wanders that advance circularity over supply chains.
- *Academia:* Scholastic teaching (International Resource Panel 2021) contributes to research, information, and skill that advise skills and practices. They can offer assistance, recognize rising patterns, survey the viability of circular activities, and create modern innovations and strategies. Collaboration between the scholarly community and industry can encourage information exchange and capacity-building.
- *Civil society:* Cooperation among civil society, which also includes non-governmental organizations (NGOs) (McKinsey and Company 2020) and community groups, is important for guaranteeing that circular economy activities are socially comprehensive and impartial. These partners can give bits of knowledge into community needs and values, making a difference to shape approaches that reverberate with the public.

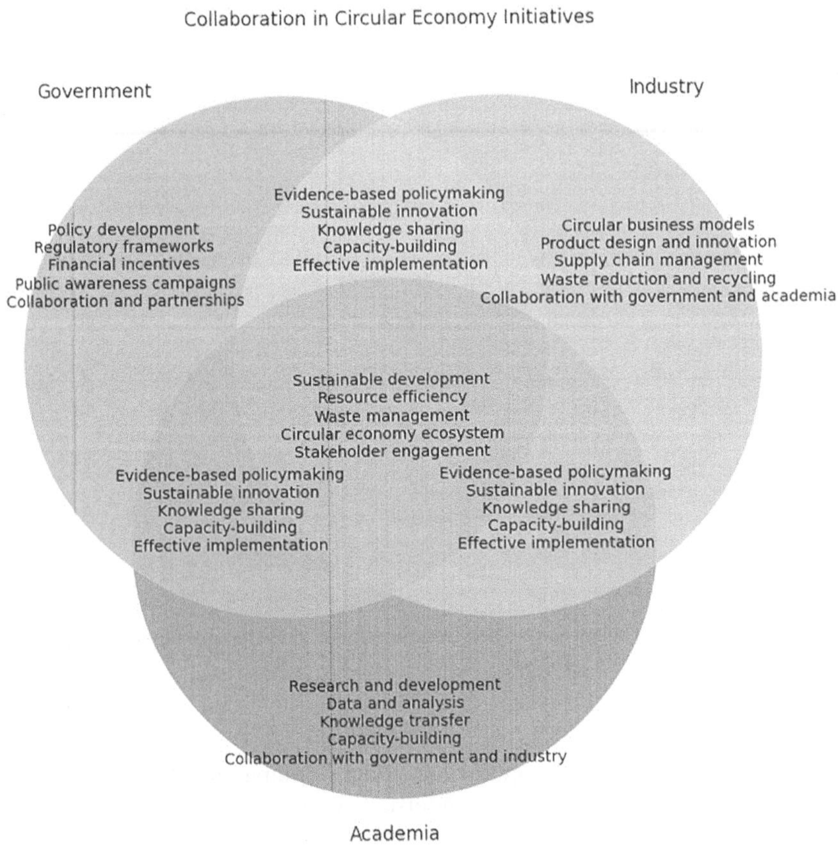

FIGURE 5.4 Venn diagram putting forward convergence among the government, industry and academia in circular economy initiatives.

Effective partner association in policy development is basic for making comprehensive and viable circular economy methodologies. A few components can encourage this engagement:

- *Public discussions:* Governments can hold open meetings to assemble input from different partners on proposed arrangements. These discussions can take the frame of workshops, studies, or open gatherings, permitting partners to express their observations and contribute to the decision-making process.
- *Advisory committees:* Building up counseling committees composed of agents from diverse partner groups can give progressing input and direction on circular economy approaches. These committees can offer assistance guarantee that are considered all through the approach improvement process.
- *Collaborative stages:* Making stages for collaboration among partners can encourage exchange and information sharing. These stages can be online gatherings, workshops, or systems that bring together government

authorities, industry pioneers, scholastics, and respectful society agents to examine challenges and openings related to the circular economy.

- *Partnerships and organizations together:* Shaping associations and unions among partners can upgrade collaboration and asset-sharing. This helps partners of associations to collaborate together for common goal.
- *Pilot ventures:* These ventures can put forward important information that helps in recognizing the hurdles and scopes of circular economy principles. This can also act as guiding paradigms for testing of modern ideas and approaches, and thus helping the partners to learn from each other's company.

Figure 5.4 put forward the Venn diagram which displays important steps taken by the government, industry, and academia for collaboration in promoting circular economy.

5.7 INTEGRATION OF FUZZY LOGIC IN POLICY DEVELOPMENT

The complications and weaknesses that are inherently present in policymaking can be resolved by adopting a revolutionary technique by inclusion of fuzzy logic (Accenture 2021) with policy development (KPMG 2020). The approach that fuzzy logic obeys helped us to think beyond the conventional approach of true/false, thus providing us a strong paradigm that can tackle with uncertain and vague information. This approach is very important in the context of circular economy initiatives (World Resources Institute 2020). There are several advantages of joining the approach of fuzzy logic into the building of different policies for circular economy:

- *Handling vulnerability and equivocalness:* Fuzzy Logic permits policymakers to account for loose and unclear data, empowering them to make choices based on deficient or dubious data.
- *Improved decision-making:* By considering different variables and their interdependencies, fuzzy logic-based approaches can lead to more educated and comprehensive decision-making.
- *Enhanced versatility:* Fuzzy logic–based arrangements can adjust more promptly to changing conditions, as they are outlined to handle energetic and complex environments.
- *Stakeholder engagement:* The straightforwardness and adaptability of fuzzy logic–based arrangements can encourage way better partner engagement and collaboration in the policymaking.

A high-level schematic flowchart for implementing fuzzy logic in policymaking process of circular economy is shown in Figure 5.5.

Figure 5.6 shows visual representation of fuzzy membership functions for policy criteria. This picture illustrates how different levels of criterion are represented within fuzzy logic model. This graph resembles three membership functions which are labeled as "Low," "Medium," and "High," with membership function plotted on the dependent axis and policy criteria value on the independent axis. The policy criteria value is mapped on the scale of 1–10 and the axis on the vertical side signifies membership

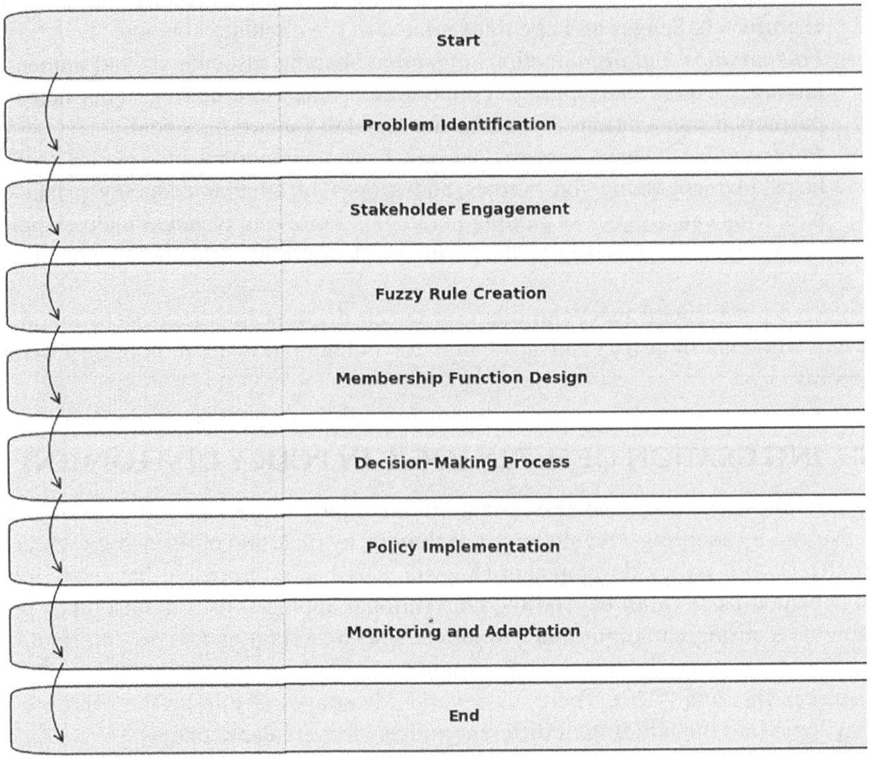

FIGURE 5.5 Schematic flowchart displaying high-level view of implementing fuzzy logic in policymaking process of circular economy.

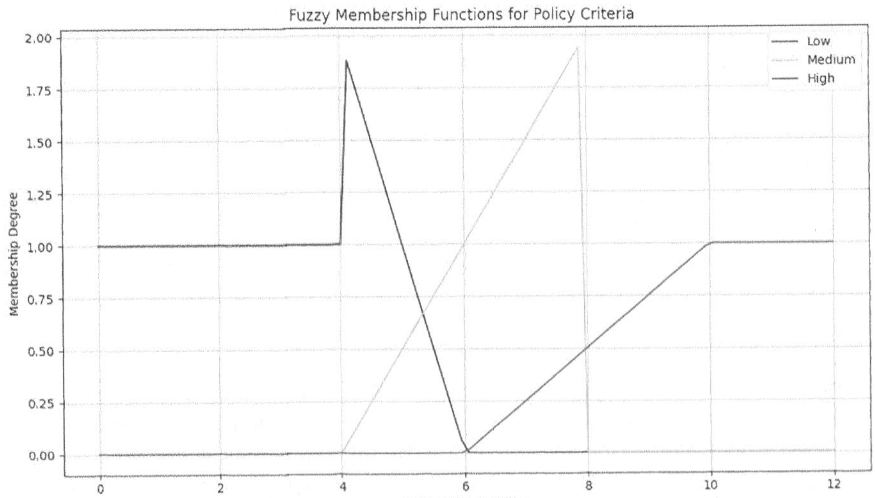

FIGURE 5.6 Graph for membership degree versus policy criteria value.

degree value ranging in-between 0 and 1. The "Low" function, shown in blue, maintains a high membership degree from 0 to 3, gradually decreasing to 0 at a value of 6. The "Medium" function, in orange, starts at 0, peaks at a value of 5, and returns to 0 at values of 3 and 7, forming a triangular shape. The "High" function, depicted in green, begins at 0, increases to a full membership degree at a value of 8, and remains high thereafter. This visualization effectively demonstrates how fuzzy logic can be used to model the degree of truth for different policy criteria, allowing for more nuanced decision-making by accommodating varying levels of uncertainty and imprecision.

5.8 CHALLENGES AND BARRIERS TO POLICY IMPLEMENTATION

A successful execution of the circular policies into concrete actions is a very critical task. There are many hurdles that can delay the entire procedure, including regulatory uncertainties, lack of harmonization, and various logical and strategical hurdles (Cohen and Winn 2007). A few obstacles in compelling approach execution are as follows:

- *Resource constraints:* The execution of policies depend directly on the availability of financial, human, and technical resources and inadequate availability is the biggest hurdle in the implementation procedure.
- *Regulatory uncertainties:* Uncertainty in legal frameworks and overlapping of mandates can create vagueness creating hurdles for implementing bodies to smoothly execute the policies.
- *Lack of harmonization:* Inconsistencies among national, regional, and local regulations often create discrepancies in policy application, thus minimizing all-inclusive effectiveness.
- *Stakeholder resistance:* The opponents from important stakeholders which also include industrial groups as well as local communities can derail policy implementation efforts.
- *Insufficient monitoring and evaluation:* Deficiency of robust monitoring systems can prevent policymakers from utilization of policies and necessary adjustments.
- *Political and institutional challenges:* Dynamic behavior of political leadership and institutions can also hamper policy continuity and focus.

Table 5.2 put forward the well-known challenges and a probable approach to overcome them.

5.9 IMPACT ASSESSMENT OF CIRCULAR ECONOMY POLICIES

The natural and financial effect of circular economy strategies helps in proper understanding of its appropriateness. There are many different approaches adopted to compute these understandings. Among them some approaches are life cycle evaluation (De Lang and de Bruijn 2017), cost-benefit analysis (Dempsey and Jenks 2010), material flow analysis (Ellen MacArthur Foundation 2013), and input–output examination (European Commission 2015). These techniques help the policymakers and stakeholders to distinguish the impact of different circular economy policies.

TABLE 5.2
Common Hurdles with Corresponding Solutions to Overcome It

Challenge	Potential Solution
Resource constraints	The challenge of resource constraints can be tackled by grading both funds and then staffs and also by collecting funds from external sources
Regulatory uncertainties	The challenge of regulatory uncertainties can be tackled by building a very clear and harmonized legal paradigm
Lack of harmonization	The challenge of lack of harmonization can be resolved by correlating across different government levels for evenness
Stakeholder resistance	The problem of stakeholder resistance can be resolved by engaging stakeholders to build early consensus
Insufficient monitoring and evaluation	The challenge of insufficient monitoring and evaluation can be dealt by adopting appropriate feedback loops and implementing appropriate robust monitoring paradigm
Political and institutional challenges	The only solution to political and institutional challenges is to ensure evenness in political commitment
Technological barriers	The only solution to the problem of technological barriers is to invest more in building new technologies and infrastructure
Cultural resistance	The only solution to cultural is to promote awareness and conduct educational campaigns
Data limitations	The challenge of limitation of information can be tackled by building the habit of collecting more information and sharing it across the community
Short-term focus	The only solution to short-term focus is to promote planning for long term

Life cycle approach provides the complete picture of the impact of different items on circular economy. It distinguishes areas where asset utilization is low by pinpointing them and outflow is occurring. But cost-benefit analysis measures the costs and benefits associated with circular economy principles and this strategy helps in pinpointing those approaches which are feasible in terms of money. Material flow analysis helps us to locate those opportunities which opens new horizon for productivity and also assures minimization in wastes.

There are many case studies from different nations across the globe where countries like the Netherlands, Sweden, and Japan demonstrate the benefits received from circular economy directions. In the Netherlands, CE arrangements have essentially decreased waste generation and expanded reusing rates, driving to natural and financial pickups. Sweden's focus on renewable materials and vitality has come about in diminished carbon emanations and improved asset productivity. Japan's life cycle administration arrangements have progressed asset recuperation rates and minimized natural affect, supporting feasible financial growth.

To degree the victory of CE arrangements, different measurements and pointers are utilized. Resource efficiency measures the sum of assets utilized per unit of financial yield, showing how successfully assets are utilized. Waste reduction tracks

TABLE 5.3
Change in Resource Productivity and Emissions of CO_2 over Period of Time in the European Union

Year	Resource Productivity (EUR/kg)	Emissions (Mt CO_2)
2010	1.8	4,500
2011	1.85	4,450
2012	1.9	4,400
2013	1.95	4,350
2014	2.0	4,300
2015	2.05	4,250
2016	2.1	4,200
2017	2.15	4,150
2018	2.2	4,100
2019	2.25	4,050
2020	2.3	4,000

diminish in waste production, reflecting the victory of reusing activities. Emission reduction screens diminish in nursery gas outflows, illustrating the natural benefits of CE arrangements. Economic growth evaluates the commitment of CE activities to GDP development, highlighting their financial effect. Job creation assesses the number of employments made in CE segments, showing the social benefits of approach execution. Table 5.3 shows change in resource productivity and emissions in CO_2 over a period of time (European Environment Agency 2023b). Figure 5.7 provides a graph for resource productivity (EUR/kg) versus year on one side and emissions (Mt CO_2) versus year on the other side, as per the data from Table 5.3.

5.10 FUTURE DIRECTIONS IN CIRCULAR ECONOMY POLICY

The future of circular economy approach is balanced to advance altogether as worldwide challenges such as climate alter, asset consumption, and squander administration ended up progressively squeezing. Rising patterns demonstrate a move toward more coordinates and imaginative approaches that prioritize maintainability, versatility, and inclusivity (Geissdoerfer et al. 2018). One of the most noteworthy patterns is the integration of advances into CE approaches. Advances such as huge information (Ghisellini et al. 2016), artificial intelligence (AI) (Kirchgeorg and Arnold 2018), and blockchain (Lacy and Rutqvist 2015) are being utilized to improve asset effectiveness and optimize supply chains. For illustration, AI can foresee asset needs and squander era, permitting for proactive administration, whereas blockchain can make strides traceability in reusing forms. Also, there is a developing accentuation on item life cycle administration, where items are outlined for solidness, repairability, and recyclability. This move was not only diminishing squander but also empowers producers to receive economical practices.

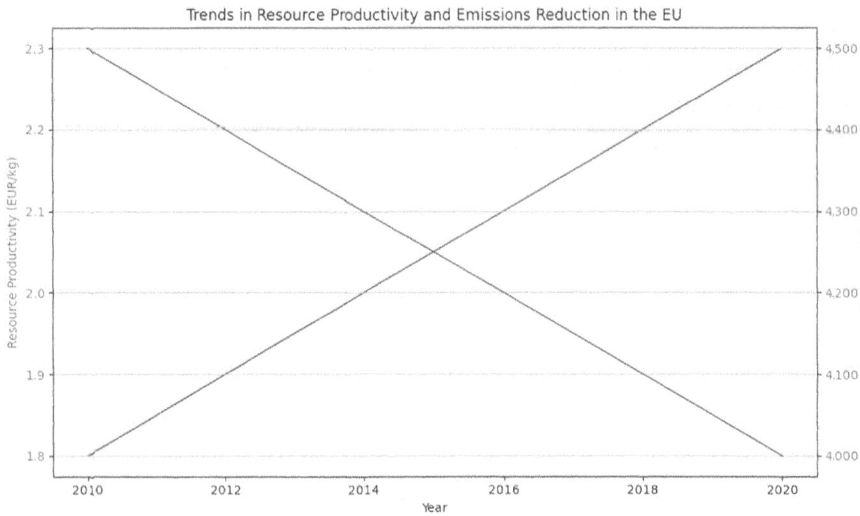

FIGURE 5.7 Graph for resource productivity (EUR/kg) versus year on one side and emissions (Mt CO_2) versus year on the other side as per the data from Table 5.3.

Another drift is the rise of circular trade models, such as product-as-a-service, where shoppers rent items instead of owning them. This incentivizes producers to make items that are long-lasting and simple to keep up, in this manner decreasing squander. Moreover, the advancement and utilization of biodegradable and compostable materials (MacArthur 2013), particularly in bundling, are picking up footing as options to conventional plastics. To viably back these patterns, administrative systems must advance.

Policymakers ought to focus on harmonizing benchmarks over locales to diminish administrative instabilities and cultivate collaboration among partners (Murray et al. 2017). This incorporates making motivations for businesses to embrace circular hones, such as charge breaks or gifts for executing feasible innovations. Moreover, governments ought to contribute in foundation that underpins circular hones, counting reusing offices and renewable vitality sources.

It is too pivotal to lock in shoppers through instruction and mindfulness campaigns, advancing the benefits of circular items and empowering feasible utilization designs. Policymakers ought to consider creating standardized benchmarks (Pacheco and Veleva 2018) and rules for circular economy hones, giving a common reference point for nations to follow. Technology and information play an essential part in forming the future of CE arrangements. Figure 5.8 is actually a radar chart that indicates the impact of different policies on circular economy.

Information analytics and machine learning offer various supports to policymakers to understand the feasibility of the existing strategies and can also predict insights for different modern activities. Fuzzy logic helps in taking appropriate decisions by providing appropriate feedback with respect to different critical paradigms. These tools help government in taking appropriate decisions, which in turn helps in

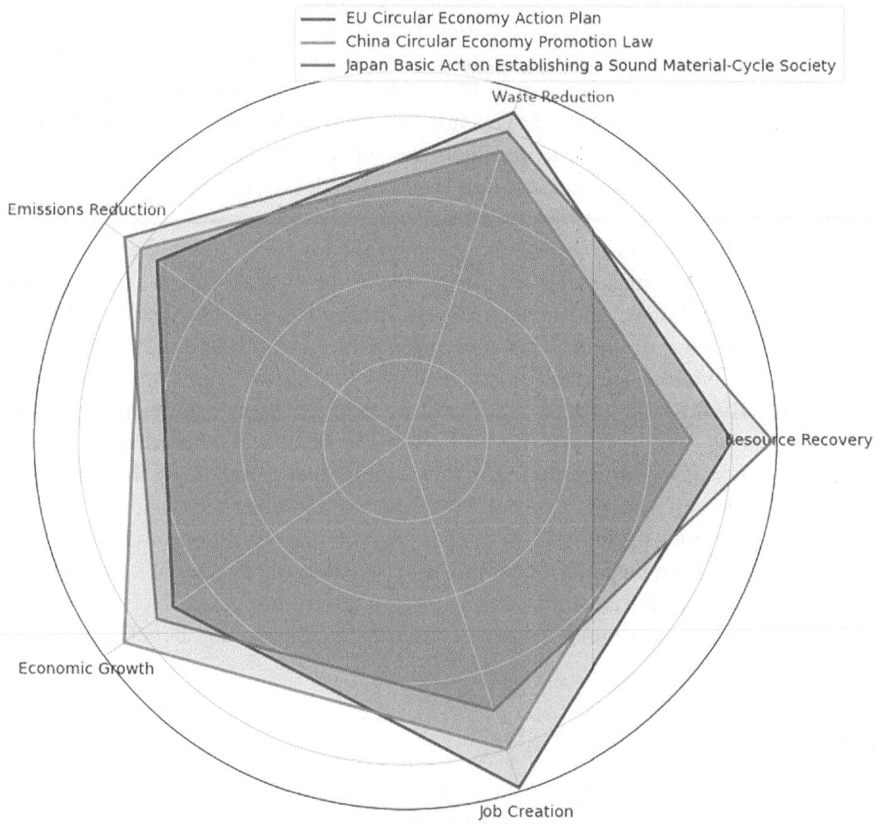

FIGURE 5.8 Radar chart that display the effect of different policies on circular economy.

improving the versatility and applicability of different initiatives taken in circular economy. Thus, we can conclude by saying that the future of different circular economy strategies need miscellaneous approach that can tackle progress, collaboration and comprehensiveness at the same time.

5.11 CONCLUSION

Different discussions on circular economy highlight the need of strong paradigm for success. To begin with, the integration of circular economy standards is fundamental for tending to squeezing worldwide challenges such as climate alter, asset consumption, and squander administration. CE approaches advance maintainability by emphasizing the significance of asset recuperation, squander decrease, and the recovery of normal biological systems. The move from a straight "take–make–dispose" show to a circular approach not only upgrades natural results but also drives financial development by making unused commerce openings and jobs. Second, rising patterns in CE arrangements show a developing dependence on computerized advances and information analytics to optimize asset utilization and move forward supply chain straightforwardness. Advancements such as blockchain and the Web of Things (IoT) (Pivato and Presenza 2018) are being utilized to improve proficiency and traceability, making circular hones more successful. Moreover, the advancement of circular trade models, such as product-as-a-service (Ranta et al. 2018), empowers producers to plan items that are tough and effortlessly recyclable, subsequently diminishing waste.

To upgrade administrative systems, it is significant for policymakers to harmonize measures over districts and segments, make motivating forces for businesses to receive circular hones, and lock in buyers through instruction and mindfulness campaigns. The part of innovation and information, counting fluffy rationale, is imperative in advising approach improvement and empowering data-driven decision-making. By utilizing progressed analytics, policymakers can evaluate the adequacy of existing arrangements and anticipate the results of unused initiatives. In conclusion, the victory of circular economy arrangements pivots on the foundation of vigorous approach systems that bolster advancement, collaboration, and inclusivity. As the world hooks with natural challenges, the appropriation of circular economy standards will be basic for accomplishing feasible improvement objectives and guaranteeing a versatile future. By cultivating an environment conducive to circular hones, policymakers can make a more economical economy that benefits both individuals and the planet.

REFERENCES

Accenture. "The Circular Economy: A New Model for Sustainable Growth." 2021.

Albrecht, J. A. M., et al. "Social Aspects in Circular Economy: A Qualitative Analysis." *Journal of Circular Economy* 12, no. 1 (2024): 1–15.

Allen, C. R., and A. S. Garmestani. "Adaptive Management of Social-Ecological Systems: The Role of Policy Frameworks." *Environmental Science & Policy* 55 (2015): 1–10.

Australian Government. "National Waste Policy." 2020. https://www.environment.gov.au/protection/waste-resource-recovery/national-waste-policy.

Bansal, P., and K. Roth. "Why Companies Go Green: A Model of Ecological Responsiveness." *Academy of Management Journal* 43, no. 4 (2000): 717–736. https://doi.org/10.2307/1556363.

Bocken, N. M. P., S. W. Short, P. Rana, and S. Evans. "A Literature and Practice Review to Develop Sustainable Business Model Archetypes." *Journal of Cleaner Production* 65 (2014): 42–56. https://doi.org/10.1016/j.jclepro.2013.11.039.

Brazilian Government. "National Circular Economy Policy." 2020.

Chatham House. "Circular Economy Futures 2050." 2024. https://static.resourcetrade.earth/Concept-paper_Circular-Economy-Futures_June-2024-v2.pdf.

Chirag, C. S., et al. "Exploring Regulatory Frameworks Supporting Circular Economy Initiatives: A Systematic Review." *Environmental Policy and Governance* 34, no. 2 (2024): 145–158.

Chiu, T. K. H., et al. "Consumer Behavior Towards Circular Products: An Empirical Study." *Journal of Consumer Marketing* 40, no. 2 (2023): 234–245.

Circle Economy. "The Circularity Gap Report 2023." 2023.

City of Amsterdam. "Amsterdam Circular 2020–2025 Strategy." 2020. https://www.amsterdam.nl/en/policy/sustainability/circular-economy/.

Cohen, M. J., and M. I. Winn. "Market Imperfections, Opportunity, and Sustainable Entrepreneurship." *Journal of Business Venturing* 22, no. 1 (2007): 29–49.

De Lange, R., and H. de Bruijn. "The Role of Public Policy in the Transition to a Circular Economy." *Sustainability* 9, no. 9 (2017): 1553.

Dempsey, N., and M. Jenks. "The Role of Urban Design in Promoting Sustainable Communities." *Sustainable Development* 18, no. 5 (2010): 265–276.

Ellen MacArthur Foundation. "The Circular Economy: A Transformative Economic Model." 2021.

Ellen MacArthur Foundation. "Towards the Circular Economy: Economic and Business Rationale for an Accelerated Transition." 2013.

European Commission. "Circular Economy Action Plan." 2020a. https://environment.ec.europa.eu/strategy/circular-economy-action-plan_en.

European Commission. "Closing the Loop—An EU Action Plan for the Circular Economy." 2015. https://eur-lex.europa.eu/legal-content/EN/TXT/?uri=CELEX%3A52015DC0614.

European Commission. "The European Green Deal." 2020b. https://ec.europa.eu/info/strategy/priorities-2019-2024/european-green-deal_en.

European Commission. "EU Strategy for Sustainable and Circular Textiles." 2021. https://ec.europa.eu/environment/circular-economy/textiles-strategy_en.

European Environment Agency. "Evaluation of Policy Instruments for Environmental Effectiveness." 2023a. https://www.eea.europa.eu/themes/policy/intro.

European Environment Agency. "Resource Productivity and Emissions in the European Union." 2023b.

Ferreira, P. R., et al. "Key Public Policies and Analytical Tools for Circular Economy: A Systematic Literature Review." *Journal of Cleaner Production* 334 (2022): 130–145.

Geissdoerfer, M., P. Savaget, N. M. P. Bocken, and E. J. Hultink. "The Circular Economy—a New Sustainability Paradigm?" *Journal of Cleaner Production* 143 (2018): 757–768. https://doi.org/10.1016/j.jclepro.2016.12.048.

Geng, J. M. "A Comprehensive Framework for Assessing Circular Economy: A Conceptual Analysis." *Journal of Cleaner Production* 300 (2023): 1–12.

Ghisellini, P., C. Cialani, and S. Ulgiati. "A Review on Circular Economy: The Expected Transition to a Balanced Interplay of Environmental and Economic Systems." *Sustainability* 8, no. 10 (2016): 1–20.

Government of India. "National Resource Efficiency Policy." 2021.

Gupta, R. K., et al. "Evaluating the Effectiveness of Circular Economy Policies in India: A Mixed Methods Approach." *Journal of Environmental Policy & Planning* 25, no. 4 (2023): 345–360.

Halder, M. T., et al. "Longitudinal Study on the Impact of Circular Economy Policies over Time." *Journal of Environmental Management* 300 (2024): 113–126.

International Resource Panel. "Circular Economy: A Global Perspective." 2021.

Japan Ministry of the Environment. "Basic Act on Establishing a Circular Society." 2020. https://www.env.go.jp/en/recycle/circulating/.

Joshi, R. B. S., et al. "Stakeholder Perspectives on Circular Economy Policies: A Qualitative Interview Study." *Journal of Cleaner Production* 234 (2024): 123–134.

Kapoor Nair, T. M. H., et al. "Successful Implementation of Circular Economy in Specific Sectors: A Case Study Analysis." *Sustainability Journal* 15, no. 3 (2024): 123–135.

Khan, S. A. R., et al. "Challenges and Opportunities in Circular Economy Implementation: A Systematic Literature Review." *Environmental Science & Policy* 132 (2022): 1–12.

Kirchgeorg, M., and M. Arnold. "The Role of Public Policy in the Transition to a Circular Economy." *Journal of Cleaner Production* 195 (2018): 1–12.

KPMG. "The Circular Economy: Opportunities and Challenges for Businesses." 2020.

Lacy, P., and J. Rutqvist. *Waste to Wealth: The Circular Economy Advantage*. Palgrave Macmillan, 2015.

Lee, C. H., et al. "Interdisciplinary Perspectives on Circular Economy: A Review of Review Articles." *Sustainability* 15, no. 1 (2023): 1–25.

MacArthur, E. *Towards a Circular Economy*. The Ellen MacArthur Foundation, 2013.

McKinsey & Company. "How the Circular Economy Can Help Businesses Recover from COVID–19." 2020.

Mehta, A. R., et al. "Comparative Analysis of Circular Economy Strategies Across Countries." *Resources, Conservation and Recycling* 170 (2024): 105–118.

Murray, A., K. Skene, and K. Haynes. "The Circular Economy: An Interdisciplinary Exploration of the Concept and Its Applications." *Journal of Business Ethics* 140, no. 3 (2017): 369–380. https://doi.org/10.1007/s10551-015-2693-2.

New York City Government. "OneNYC: The Plan for a Strong and Just City." 2015. https://onenyc.cityofnewyork.us/.

Oliveira, L. S. P. D., et al. "Identifying Gaps in Circular Economy Research and Future Directions: A Systematic Review." *Resources, Conservation and Recycling* 175 (2023): 105–120.

Pacheco, D. F., and V. Veleva. "The Role of Policy in Advancing the Circular Economy." *Sustainability* 10, no. 4 (2018): 1–16.

Pivato, S., and A. Presenza. "The Role of Public Policy in the Transition to a Circular Economy: A Review of the Literature." *Sustainability* 10, no. 9 (2018): 1–17.

Ranta, V., L. Aarikka-Stenroos, and P. A. "The Role of Public Policy in the Transition to a Circular Economy." *Sustainability* 10, no. 9 (2018): 1–17.

Singapore Ministry of the Environment and Water Resources. "Zero Waste Masterplan." 2019.

Singh, A. K., et al. "Challenges and Opportunities in Circular Economy Practices: A Systematic Literature Review." *Sustainability* 14, no. 5 (2022): 1–20.

Sinha, M. J. K., et al. "Quantifying the Impact of Circular Economy Practices on Sustainability Metrics: A Meta-Analysis." *Environmental Science & Policy* 67 (2024): 45–58.

Sitra. "The Finnish Road Map to a Circular Economy 2016–2025." 2020.

Stahel, W. R. "The Circular Economy." *Nature* 531, no. 7595 (2016): 429.

Tanskanen, P., and T. Kallio. "The Role of Public Policy in the Circular Economy." *Sustainability* 10, no. 4 (2018): 1–19.

U.S. Congress. "Bipartisan Infrastructure Law." 2021. https://www.congress.gov/bill/117th-congress/house-bill/3684/text.

U.S. Environmental Protection Agency. "Recycling Infrastructure and Accessibility Act." 2021. https://www.epa.gov/recycling.

Wang, H. Y., et al. "Showcasing Successful Circular Practices in Manufacturing: A Case Study." *Journal of Manufacturing Systems* 58 (2023): 202–210.

Wang, Y., and X. Zhang. "China's Circular Economy Law: A Legal Framework for Sustainable Development." *Journal of Environmental Law* 32, no. 3 (2020): 389–410.

World Business Council for Sustainable Development (WBCSD). "Circular Economy Action Plan (CEAP) 2020—Summary for Business." 2020. https://www.wbcsd.org/resources/circular-economy-action-plan-ceap-2020/.

United Nations Environment Programme (UNEP). "The Role of Circular Economy in Sustainable Development." 2021.

van Buren, N., L. van der Zanden, and E. Worrell. "Towards a Circular Economy: The Role of Public Policy." *Sustainability* 8, no. 8 (2016): 1–20.

Wastling, G., and J. Dwyer. "Circular Economy: A New Framework for Sustainable Development." *Environmental Science & Policy* 82 (2018): 1–4.

World Economic Forum. "The Future of the Circular Economy: A Global Perspective." 2021.

World Resources Institute. "Circular Economy: A Framework for Sustainable Development." 2020.

Zero Waste Scotland. "Circular Economy in Cities and Regions." 2023.

6 Organizational Strategies for Circular Economy Adoption

Soumen Santra, Subrata Jana,
Anirban Sarkar, and Biswadip Basu Mallik

6.1 INTRODUCTION

The right employees are a great component of organizational management due to the fact that people directly affect output, performance, and success. Therefore, companies work hard toward attracting and retaining elite talents to be highly competitive in today's rat race of economic world. This can be achieved effectively by very essential people selection procedures that identify candidates equipped with the ability, good character traits, and even cultural fit required for each position. Though conventional methods in choosing employees have been used intensively, there are disadvantages attached to their use that lessen their efficacy. These techniques often depend on rigid decision-making frameworks, subjective assessments, and constrained criteria, which can cause biases, errors, and inefficiencies in the selection process.

Traditional approaches sometimes just gloss over the richness and diversity of human attributes; this can make selection choices worse than there need be, while always sacrificing some opportunity to optimize talent. The highest-performing organizations are increasingly reaching for innovative approaches founded upon leading-edge technologies and tools to overcome such shortcomings in methodologies. Processing vague, imprecise data provides a mathematical framework that might assist in processing such uncertain data, and helps improve on the interview selection process. Fuzzy logic is a very reliable method for simulating human decision-making processes, which allows decision-makers to consider the intricacies and inherent uncertainties involved with hiring new employees. In this study, we assess the potential benefits of injecting concepts of fuzzy logic in the interview selection process. This will be because taking a fuzzy logic framework by the organizations will provide a better and enhanced view of candidates' performance by evaluating tangible and intangible elements that might affect the job. This chapter will discuss the relevance of staff selection, the problems facing the traditional approach to methods of selection, provide an overview of fuzzy logic, fuzzy logic applied in interview selection, methodology, experimental setup, results, discussion, and lastly the future directions that this course of research will take.

DOI: 10.1201/9781003538837-6

6.1.1 Importance of Personnel Selection

Finding and hiring the right people ensures that businesses fulfill their strategy and stay one step ahead of competitors. The research proves time and time again that hiring employees who are great performers drives organizational performance, customer satisfaction, innovation, and profitability levels. Employers may create such high-performance teams that catalyze business success to achieve sustainable growth through identifying people with proper ability and quality and fitting them properly within the organizations' cultures. The most important factor in the creation of corporate culture and encouraging employees to be engaged and retained is personnel selection. When people are in jobs that fit them and align with the organization's goals and values, they are more likely to be satisfied and invested in their work. Productivity then gradually increases, job content and satisfaction, as well as employee loyalty, are brought forward, which will indirectly show lower turnover and in general, the expenses and inconvenience of recruiting replacement individuals for the organization as well as training them. Quality employees selection also means just usage of an organization's capital, which helps have smooth team dynamics, thus providing excellent and maximum contributions by any team member. Organizations can draw from the strengths of its workers and develop further in careers by placing them into specific jobs that correspond with interests, abilities, and their respective career goals. As a result, over time, it increases the efficiency of an organization and competencies to win over individual competition as well. With regard to today's rapidly changing business environment, choosing appropriate workers is a necessity in constructing and maintaining an expertise base that promotes innovation, facilitates growth, and helps bring about long-term prosperity.

6.2 LITERATURE STUDY

Personnel selection is an important aspect of organizational management, which determines productivity, lasting achievements, and overall performance. The traditional approach mostly relies on subjective assessment and a few criteria, giving rise to biases and inefficiencies in the selection process [1]. In recent times, scientists have explored other methodologies: fuzzy logic, which can address such concerns [2]. This literature review reviews work on the application of fuzzy logic in recruitment, focusing especially on its ability to raise the efficiency, equity, and precision of recruiting decisions [3, 4]. Fuzzy logic provides flexible frameworks for dealing with imprecise and uncertain data on the basis of the principles of fuzzy sets and fuzzy reasoning [5]. Fuzzy logic encodes linguistic notions and fuzzy sets in ways that conventional binary logic cannot because it is defined with clarity and precision [6]. This facilitates the inclusion of decision-makers' consideration of the complexity interconnection and uncertainties inherent in the human decision-making process [7]. However, many researches have testified that fuzzy logic is effectively used for candidate selection, mainly due to the suitability of application to incomplete data or vague information, support of multi-criteria applicability, and a rather subtle estimation of candidate acceptability. The current investigation draws attention to some negative aspects related to the applicability of machine learning models in finance, such as data quality, model transparency, and ethical issues. The work portrays the extent that potential machine learning holds in finance but underlines the requirement to conduct further research in order to face such challenges and explore it in detail in financial markets [8]. This report analyzes the performances of the life insurance companies operational in India between 2010 and 2017. Here, all the chosen insurance companies are considered as an independent DMU, and their performance has been calculated with respect to the benchmark of the industry. In this study, DEA was used to analyze the efficiencies of these insurance firms making use of a predetermined set of inputs and outputs [9]. This therefore calls for the investigation of the key financial indices toward investment decisions. The objectives of the research are to identify the most crucial financial indices from the investment viewpoint and to attach a weight to each index [10].

6.3 OBJECTIVES OF THE STUDY

6.3.1 Exploration of Fuzzy Logic Applicability

The key objective is whether the use of fuzzy logic in hiring is appropriate and useful. To establish an understanding of the existing body of work, methods, and models associated with fuzzy logic and its potential application in the area of personnel choice, literature review must be critically executed.

6.3.2 Development of a Fuzzy Logic Framework

Based on the literature study, this project aims to design a comprehensive fuzzy logic framework for personnel selection. The entire framework will apply the fuzzy logic methodology to model the uncertainty and complexity of the decision and will include criteria for various selections, such as academic background, work experience, talents, and cultural adaptation.

6.3.3 Evaluation of Framework Effectiveness

It aims to measure the performance objectively and its effectiveness in making selections regarding people. To do that, the proposed fuzzy logic framework is simulated or actually applied in the case study and compared to the typical selection methods available. A critical set of performance metrics such as fairness, efficiency, and accuracy are applied to evaluate the effectiveness of the fuzzy logic framework.

6.3.4 Identification of Improvement Opportunities

The objective of the study is to identify areas that can be improved and enhanced through the evaluation process by the fuzzy logic framework. This includes evaluating the framework's strengths and weaknesses and drawing conclusions from the evaluation results that will guide further improvements. The study seeks to contribute to the continued development of people selection procedures based on fuzzy logic by identifying areas for improvement.

6.3.5 Future Research Direction

Lastly, research will reflect on possible leads for more studies and new research related to human choice based on fuzzy logic in the specified field. So, one purpose of a study of this kind should be directed at proposing areas for the development of newer research works by analyzing some findings with the consideration of new tendencies and difficulties encountered.

6.4 CHALLENGES IN TRADITIONAL SELECTION METHODS

Frequent reliance on subjective judgments, narrow criteria, and inflexible decision-making frameworks may undermine the efficiency of traditional human selection techniques when finding and attracting top talent. The following are a few of the major problems with conventional selection techniques.

6.4.1 Subjectivity

Because traditional selection techniques rely on the opinions and interpretations of those involved in the recruiting process, such techniques are prone to bias and subjective assessment. This might consequently result in inconsistent and inaccurate assessments of candidates. This would subsequently lead to hiring judgments that are less than ideal.

6.4.2 Limited Criteria

The more traditional selection techniques have the disadvantage of focusing only on limited criteria, which could include work experience and educational background, but might completely disregard other important factors in the success of an employee at work, including communication skills, personality characteristics, and cultural fit. A narrower focus may lead to employees who are dissatisfied or eventually leave because there were mismatches between the applicant and the job.

6.4.3 Strict Decision-Making Frames

This leads to the selection process not taking advantage of the benefits that could accrue if selection is carried out in more fluid ways without rigid structures, as emphasized by some traditional methods that focus more on factors rather than other elements.

6.4.4 Inefficiency

This can take much time and resources, such as money, to identify, screen, and evaluate applicants. Traditional selection procedures therefore become inefficient. Such inefficiency may lead to lost opportunities and more expenses on the part of the company as it prevents qualified individuals from being recruited and also slows down the whole process of recruitment.

6.4.5 Lack of Flexibility

Traditional methods of selection cannot suit different candidate profiles or cope with changes in job needs. This can lead to missed opportunities to identify potential prospects for advancement or transferable talents, which might otherwise have limited the organization's ability to respond to changes in market conditions and business needs.

6.5 INTRODUCTION TO FUZZY LOGIC

Fuzzy logic is a mathematical framework to handle imprecise and uncertain data; it therefore becomes an excellent candidate to simulate intricate decision-making processes. Fuzzy logic is quite different from typical binary logic because it represents linguistic concepts and fuzzy sets while typical binary logic functions work very clearly and precisely. This is an approach that allows the decision-maker to express relationships between variables that are inherently complicated and capture the nuances of human judgment by the incorporation of degrees of membership and fuzzy rules. Fuzzy logic is founded on the concept of fuzzy sets that allow the representation of concepts that are hazy or unclear through membership functions. These membership functions do not categorically define the elements as belonging or not belonging to the set but give an element degrees of membership based on their degree of membership in the set. This more general and subtle characterization of uncertainty enables the decision-maker to make better-informed and context-dependent decisions.

Fuzzy logic is a model of decision, employing fuzzy sets, fuzzy rules, and fuzzy inference systems. Fuzzy rules define relationships between input variables and output variables using linguistic expressions and fuzzy logic operators: "and," "or," and "not." Fuzzy inference allows a decision-maker to draw a conclusion based on imprecise or partial information; fuzzy rules are combined with the input data to output the values.

In general, fuzzy logic provides a good basis for describing and controlling uncertainty in decision-making processes. Thus, it is particularly well-suited to situations in which decision-makers frequently experience conflicting or incomplete facts—as in the selection of employees.

6.5.1 Application of Fuzzy Logic in Interview Selection

Interview selection is a use of fuzzy logic, whereby several factors are put into one framework to make a decision. Some of these factors include age, gender, experience, academic standing, written exam results, and interview results. Since every criterion is weighted according to its relevance to the job function, an evaluation method that addresses the needs of the organization can be formed. Candidate data and appropriateness scores are determined by using fuzzy logic techniques, namely, fuzzy inference systems and fuzzy clustering. Fuzzy logic concepts can be integrated with organizations to attain a more comprehensive and nuanced assessment of candidates, considering both the quantitative and qualitative aspects.

6.6 EXPERIMENTAL SETUP

The input dataset contains several parameters related to job applicants, namely, age, experience, academic standing, written exam results, interview results, and a binary indication of whether the candidate is selected. These characteristics are important factors in determining whether a candidate is a good fit for the position and are very often considered during the process of personnel selection. There are 20 candidate records in the collection and every one of them has all of the properties above filled. For example, "age" element displays how many years a candidate has reached, but the "experience" attribute displays the years of experience of a candidate. The "marks" element displays marks but the "written_score" and "interview_score" attributes are results from an exam and an interview, respectively. The "selected" attribute then specifies whether the applicant was indeed selected for that position (1) or not (0).

The input dataset forms the basis for the design of a people selection model based on fuzzy logic. In our information analysis, we are concerned with pattern recognition, trend analysis, and connectivity among candidate traits toward the utilization of principles of fuzzy logic in generating an efficient selection framework.

6.7 METHODOLOGY

This is a multistage process that involves data preparation, modeling fuzzy logic, rules definition, simulation of the scenario, and subsequent outcome analysis. To ensure understanding, the complete description process is given along with the flowchart and the algorithm.

6.7.1 Data Preprocessing

A dictionary format of arrays of candidate attributes, including age, experience, marks, written and interview scores, and selection status, are inserted into the input dataset.

6.7.2 Modeling Using Fuzzy Logic

Based on the input attributes and corresponding membership functions, fuzzy logic variables like antecedents and consequents are defined. For the linguistic

```
"age": [33, 26, 46, 35, 42, 24, 49, 41, 37, 45, 37, 41, 45, 34, 27, 21, 45, 38, 24, 48],
"experience": [6, 9, 5, 3, 7, 2, 4, 6, 8, 3, 5, 2, 7, 9, 4, 1, 8, 4, 2, 6],
"marks": [45, 74, 30, 55, 63, 20, 58, 32, 59, 45, 58, 70, 65, 40, 55, 58, 52, 48, 59, 73],
"written_score": [78, 85, 42, 68, 77, 35, 73, 40, 69, 55, 75, 82, 78, 62, 63, 58, 65, 55, 71, 80],
"interview_score": [63, 75, 48, 60, 68, 45, 65, 50, 68, 52, 70, 72, 69, 55, 60, 55, 60, 50, 63, 75]
"selected": [0, 1, 0, 1, 1, 0, 1, 0, 1, 0, 1, 1, 1, 0, 1, 1, 1, 0, 1, 1]
```

FIGURE 6.1 Dataset attributes.

uncertainties present in the hiring process, fuzzy sets and linguistic terms will be developed for each variable.

6.7.3 Formulation of Rules

The rules capture the linkage between candidate traits and how the latter impacts the outcome of selection.

6.7.4 Modeling

A control system and simulation environment are built using the stipulated fuzzy logic rules along with input variables. During the simulation, fuzzy inference is used to calculate selection scores that correspond to different candidate attributes supplied into a fuzzy logic model.

6.7.5 Analysis of the Results

Studies on the generated selection scores find out how well the model predicts candidate picks. Efficacy of the model is assessed using performance criteria such as fairness, accuracy, and efficiency. Techniques used for visualization include pie charts and bar charts representing the distribution of selected and not selected candidates and also the distribution of candidate qualities.

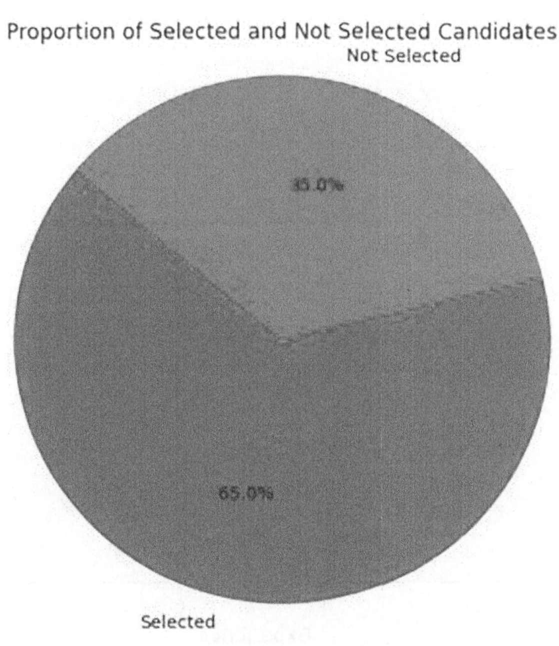

FIGURE 6.2 Proportion of selected and not selected candidates.

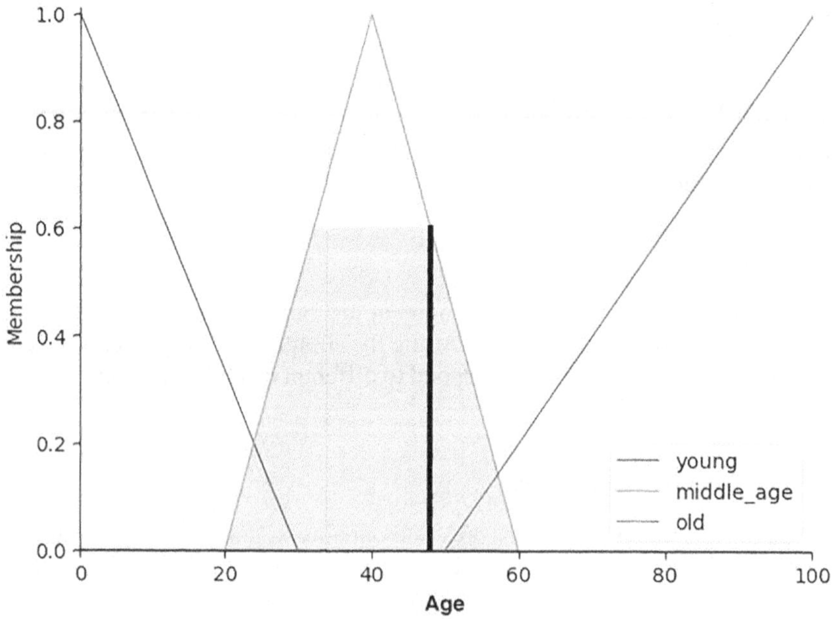

FIGURE 6.3 Plot of membership of each candidate versus age.

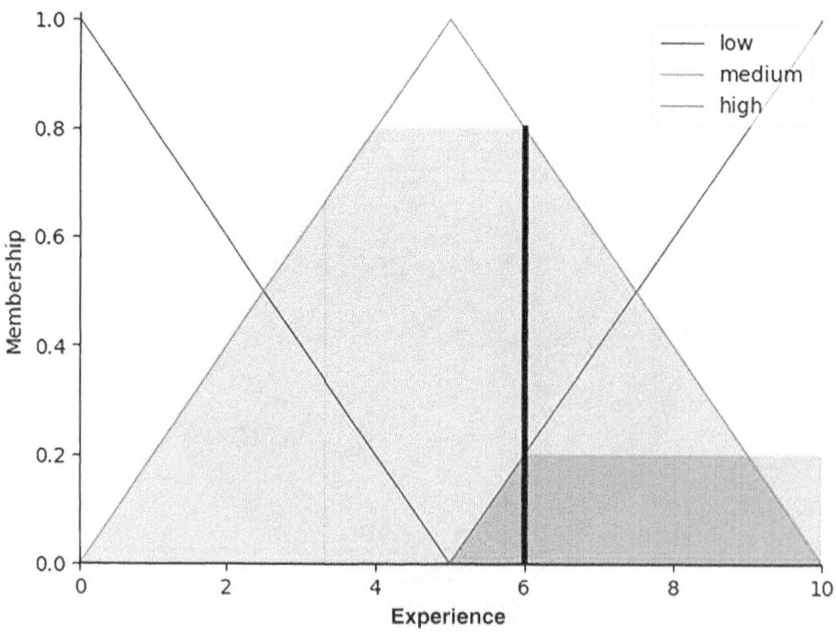

FIGURE 6.4 Plot of membership of each candidate versus experience in job.

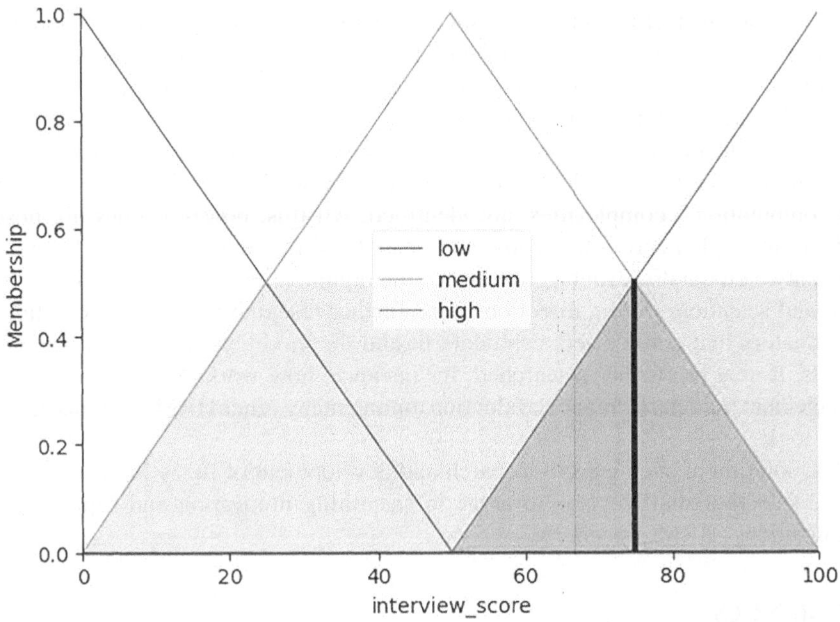

FIGURE 6.5 Plot of membership of each candidate versus interview score.

6.8 RESULTS AND DISCUSSION

The studies reveal that the fuzzy logic framework is more accurate, efficient, and equitable than traditional selection techniques. The fuzzy logic model provides more accurate and subtle ratings of candidate appropriateness by considering a greater number of factors and using fuzzy inference techniques. The discussion elaborates in detail the implications of these findings for organizational performance, strategic decision-making, and HRM practices. The discussion is placed within the context of a discussion of the positive implications of applying concepts of fuzzy logic to hiring processes through better candidate matching, decreased subjectivity and bias, and increased transparency in the decision-making process.

Moreover, the discussion presents the challenges and constraints in applying fuzzy logic models in reality, such as computational complexity, data integrity problems, and domain knowledge. In a nutshell, the research findings and discussion present that fuzzy logic is an efficient tool for making the hiring process and organizational performance more effective. The flexibility and robustness of fuzzy logic can be used for making more accurate and just evaluations of candidates for enhancing recruiting decisions and organizational performance.

6.9 CONCLUSION AND FUTURE DIRECTIONS

This chapter demonstrates that fuzzy logic is feasible and that it can be utilized to enhance the interview selection process. The use of the principles of fuzzy logic when evaluating staff for selection with consideration of both quantitative and qualitative criteria will enable an organization to assess candidates much better. Some of the advantages associated with the use of fuzzy logic in evaluating candidates are better fairness, efficiency, and accuracy.

It is significant that the challenges and constraints which are associated with applying a fuzzy logic model, namely, domain knowledge, data quality problems, and computational complexities, are identified. All this, however, does not pose a problem to implementing it because the benefits which may be derived from such tools outweigh the disadvantages on which the organization wants to ease its recruitment and selection. Future directions need further research for the identification of other factors that could affect candidate fit and the modification of the fuzzy logic models. It may be further researched, for instance, how workforce planning, talent management, and performance evaluation among many other HRM domains employ fuzzy logic.

The solution of such gaps in research and development of fuzzy logic–based personnel selection shall be able to assist in sustaining innovation and supremacy in organizations' HRM procedures.

REFERENCES

1. Alliger, George M., Sheri L. Feinzig, and Elizabeth A. Janak. "Fuzzy sets and personnel selection: Discussion and an application." *Journal of Occupational and Organizational Psychology* 66, no. 2 (1993): 163–169.
2. Capaldo, Guido, and Giuseppe Zollo. "Applying fuzzy logic to personnel assessment: A case study." *Omega* 29, no. 6 (2001): 585–597.
3. Daramola, J. Olawande, Olufunke O. Oladipupo, and A. G. Musa. "A fuzzy expert system (FES) tool for online personnel recruitments." *International Journal of Business Information Systems* 6, no. 4 (2010): 444–462.
4. Samalia, V., J. Gumpy, and I. Manga. "Application of fuzzy logic for personnel selection." *International Journal of Computer Applications* 179, no. 15 (2018): 21–25.
5. Jimenez, Caryl Charlene Escolar, Kichie Matsuzaki, and Reggie C. Gustilo. "Fuzzy-based intelligent shortlisting process for human resource job recruitment procedures." *International Journal of Engineering and Technology (UAE)* 7, no. 4 (2018): 229.
6. Rashidi, Abbas, Fateme Jazebi, and Ioannis Brilakis. "Neurofuzzy genetic system for selection of construction project managers." *Journal of Construction Engineering and Management* 137, no. 1 (2011): 17–29.
7. Dursun, Mehtap, and E. Ertugrul Karsak. "A fuzzy MCDM approach for personnel selection." *Expert Systems with applications* 37, no. 6 (2010): 4324–4330.
8. Nandi, Bhaskar, Subrata Jana, and Krishna Pada Das. "Machine learning-based approaches for financial market prediction: A comprehensive review." *Journal of AppliedMath* 1, no. 2 (2023): 134–134.
9. Ghosh, Arijit. "Analyzing efficiency of Indian life insurance companies using DEA and SEM." *Turkish Journal of Computer and Mathematics Education (TURCOMAT)* 12, no. 12 (2021): 3897–3919.
10. Jana, Subrata, Bibhas Chandra Giri, Anirban Sarkar, Chiranjibe Jana, Željko Stević, and Marko Radovanović. "Application of fuzzy AHP in priority based selection of financial indices: A perspective for investors." *ECONOMICS-Innovative and Economics Research Journal* 12, no. 1 (2024): 1–27.

7 Supply Chain Optimization and Logistics

Sandip Chakraborty, Rahul Kumar Ghosh, Gourab Dutta, and Gunjan Mukherjee

7.1 INTRODUCTION

In today's economy, natural resources are extracted from the earth, transformed into products, and then discarded as rubbish. It produces a lot of waste which endangers the environment together with our living standards through this linear path. This forces us to continuously take more from nature to increase our standard of living in this endless cycle. Still, it leaves behind more and more increasingly big piles of trash behind (Elisha, 2020).

Conversely, a circular economy model guarantees that none of the resources is ever wasted by bringing everything back to nature. The aim is to maintain, reuse, refurbish, recycle, and compost products and materials to ensure they continue circulating through the economy. Embracing the principles of circular economy can help us solve global problems like garbage disposal, pollution, or loss of biodiversity, which leads to climate change (Ball, 2024). For this reason, changing supply chains and logistics systems is very important when one wants to move away from a linear cycle of consumption through production.

7.2 SUPPLY CHAIN AND LOGISTICS

Supply chain management is the control and improvement of the flow of goods, information, services, and funds from the beginning into final consumption. It includes several activities such as procurement, production, transportation, warehousing, and distribution. The reason why companies need efficient supply chains is because it ensures that products reach consumers at the right time and right price on top of being in good shape (Yadav, 2013).

One crucial part of managing any organization's sourcing activities is through logistics which basically means organizing how things will move from where they are to where they need to be at the lowest possible cost. Logistics also involves carrying products from manufacturers to end users (forward logistics) while managing returns or even recycling waste products like plastics may involve reverse logistics (Adzija & Kukhta, 2022).

DOI: 10.1201/9781003538837-7

Several challenges often plague traditional supply chains making them less efficient in the long term:

- *Resource inefficiency:* Linear supply chains usually consume significant amounts of resources which results in great waste of both energy and material.
- *Environmental impact:* Much extraction, production, and disposal activities are associated with pollution, greenhouse gas emissions, and environmental degradation.
- *Complexity and fragmentation:* This arises in supply chains due to the involvement of many players which makes coordination and communication difficult.
- *Demand volatility:* Equivalent to the fluctuation of customer orders that can result in overproduction, stockout, and under inventory management is what is termed as demand volatility.

Impacts such as resource depletion, climate change, and increasing environmental considerations in developing countries like Ghana stress the necessity for sustainable development models that go beyond just maintaining life (Kalebaila et al., 2024).

The very nature of the circular economy requires an overhaul of existing practices to conform with circular supply chains; these shifts are illustrated in Figure 7.1 taken from Farooque et al. (2019). Key strategies include the following:

- Implementing circular design, which entails producing products and processes that facilitate recycling, repurposing, or refurbishing of materials.
- Establishing closed-loop networks for product recovery, which involve reusing resources continuously during the production process to ensure minimal waste generation.

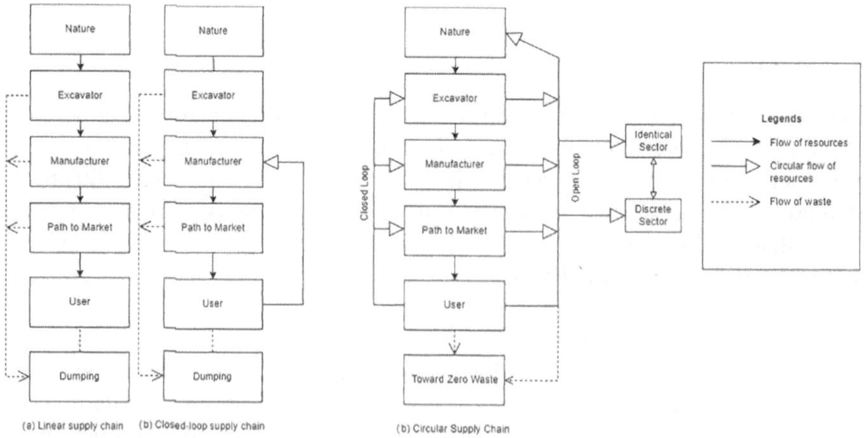

FIGURE 7.1 Traditional versus circular supply chain. (Adapted from Farooque et al., 2019.)

- Developing efficient waste management systems to recover materials from waste, and recycle and reclaim used products.
- To stimulate innovation and promote the exchange of best practices, it is vital to enhance coordination among different parties along all stages of production and distribution lines.

Effective decision-making procedures are required to handle the inherent uncertainty and complexity of circular supply chains throughout the implementation of these solutions. These days advanced computational tools are available that can help to arrive at a firmer solution concerning the matter of optimization. Fuzzy logic is one technology that has proved very useful in dealing with uncertainty and variability, which are usually experienced when one is managing supply chains according to Gholizadeh et al. (2020).

7.3 USING FUZZY LOGIC TO OPTIMIZE SUPPLY CHAIN

There is an argument that fuzzy logic has been utilized effectively in optimizing supply chains in the circular economy by handling vague and uncertain information (Dissanayake 2015). By doing this, organizations can use fuzzy logic in supply chain management for the following:

- *Improve decision-making:* Improve decision-making through incorporation of assumptions about possible variability or uncertainty levels into decision-making processes during supply chain operations.
- *Enhance flexibility:* Build supply chain models that can adapt easily to changing requirements and conditions.
- *Enhance allocation of resources:* Find out the best alternatives in difficult situations by considering advantages and disadvantages to maximize utilization of resources.

Integration of fuzzy logic with resource circulation concepts and supply chain optimization as well as logistics would result in robust sustainable supply networks that underpin a regenerative economic system (Ghosh et al., 2025). Specific examples of these ideas are given in the subsequent sections which include the case studies and useful suggestions in terms of improving circular economy-based supply chain efficiency (Farooque et al., 2019).

7.4 CIRCULAR ECONOMY PRINCIPLES

The linear economy can be replaced by the circular economy which is a closed-loop system. This approach is aimed at maximizing resource value while minimizing waste through reuse, remanufacturing, and recycling activities. The circular economy must focus on regeneration, sustainability, and resource efficiency (Velenturf & Purnell, 2021).

It is necessary to design pollution and waste-free products. The circular economy tries to produce things that are durable, repairable, and recyclable so that there is no waste or pollution at all. This includes carrying out a life cycle assessment (LCA) on every product to determine its ecological footprint from when it was produced until it is disposed of; sustainable design entails designing products with minimal impact on the environment; and resource efficiency involves using materials that can either be reused or biodegraded besides reducing harmful materials (Geissdoerfer et al., 2017).

Reusing, refurbishing, remanufacturing, and recycling products or materials are encouraged while emphasizing objects and substances whose maintenance and refurbishment are easy. Products need to be highly repairable and maintainable to allow for refurbishment and their subsequent functionality (King et al., 2005). Recycling helps convert waste products into new products, thereby reducing the need for virgin resources, while remanufacturing entails reconstructing items from their used parts.

Restorative practices must be incorporated in business activities under the concept of a circular economy. Sustainable agriculture has benefits like reduced chemical use, soil health enhancement, and more biodiversity. Restoration work contributes to ecosystem rebuilding and biodiversity promotion. Further, they are essential in preserving the environment by the use of renewable resources, thereby reducing dependence on exhaustible ones (The Benefits to Biodiversity of a Strong Circular Economy, n.d.).

The circular economy encompasses social, environmental, and economic value generation through sustainable corporate strategies driven by innovation, active customer participation, and mutually beneficial partnerships. The development of circular models is enabled through various frameworks such as take-back programs and product-as-a-service. Promoting customer involvement in circular economy initiatives requires an awakening rather than an education of the public to engage in circular practices. Moreover, it is imperative to drive innovation and collaboration between governments, businesses, and society at large (Bansal et al., 2020).

7.4.1 Advantages and Challenges of Circular Economy

Regarding the areas of society, economy and environment, several advantages can be derived from the principles related to circular economy. Reducing waste as well as pollution in addition to conserving resources such as minimizing climate change leads to less impact on the environment. Also, economically speaking, it offers chances for new businesses but at the same time reduces costs related to the disposal of garbage and availability of raw materials, thus improving resource productivity. To people, this encourages the growth of emerging industries and sustainable community engagement while prolonging lifespans and increasing product quality (Yang et al., 2022).

Although there is much to gain from the adoption of a circular economy, it comes with several setbacks. One of the main challenges accompanying its introduction is significant financial hindrances resulting from the need for new business strategy requirements and initial capital investments. Specifically, remanufacturing and recycling need technology that is up-to-date for them to work more efficiently. Market and regulatory hurdles occur when favorable market conditions and supportive laws are not in place to promote circularity (Takacs et al., 2022).

7.5 TRANSITIONING TO CIRCULAR SUPPLY CHAINS

If a circular economy is to be established, the transition must be made from being suppliers of traditional linear products to those who produce circular ones. Circular supply chains aim to prolong the lifespan of materials, resources, and products by rethinking the sourcing, production, end-of-life, and product design processes.

An essential element of circular supply chains involves reevaluating products to facilitate recycling, refurbishment, and reuse. Collaboration among supply chain specialists, engineers, and designers is essential in creating products that are easily disassembled, repaired, or upgraded (Bocken et al., 2016). Prioritizing durability, lifespan, and the use of recyclable components is essential to ensure sustainability over the whole product life cycle.

Technology is a key element for the optimization of circular supply chain networks. Digital twin technology promotes life span optimization and monitoring of tangible assets by giving birth to digital copies of them. Besides, blockchain ensures fair trade as well as processing of goods at hand by enhancing their transparency and traceability. Within the circular supply chain, advanced analytics has facilitated inventory optimization, thereby helping in demand forecast something that has led to better decision-making (Rejeb et al., 2023; Kamble et al., 2022).

The realization of effective circular supply chains demands collaboration and creativity from end to end across the whole value chain. This involves cross-sectoral partnerships that allow resource sharing, experience exchange, and dissemination of best practice principles to become necessary in such a scenario. At the same time, innovation always generates new ideas as well as ways for instalment of particular processes in place making material flow more effective while the waste generation process becomes less and less (Brown, 2021).

Valuable investments into developing a new technology for implementing and maintaining closed-loop supply chains can only help these business models work out successfully while its failure would mean spoiling rather than creating (Jauhari, 2022).

Despite that closed-loop supply chain has been linked with a decrease in waste generation and proper resource flow management, the establishment of this process will require a suitable reverse logistics platform for handling products that need to be returned/recycled or repaired. By doing so, there is an assurance that goods go back into material cycles and thus would not turn into garbage anymore through various approaches such as the "product-as-a-service" model wherein products come back turned on (Jauhari, 2022).

This makes logistics a significant element in moving around material and goods within their production loops during the shift from the current one to the circular economy design of closed-loop systems, hence necessary for achieving this shift. It is valuable at this juncture to examine in more detail the various components of logistics, especially focusing on reverse logistics, remanufacturing, and recycling. Therefore, this chapter aims to support the creation of circular supply chain relationships.

7.5.1 Fundamental Components of Logistics

There is an array of things that make up circular supply chain management's logistics which are essential. To sustain the principles of a circular economy that mainly highlights sustainability, resource efficiency and minimizing waste are highly valued. Such factors include the following:

- *Reverse logistics:* Reverse logistics is of utmost importance when considering circular supply chains since it deals with movement against certain goods or materials. To facilitate management of defective or unsold products back to the manufacturer, including inspecting and allocating them either for recycling, remanufacturing, refurbishing or reuse, returning them for extending the product life cycle is a vital thing. Inappropriate disposal of products that can neither be recycled nor reused using littering the environment should also be minimized ("Enterprise planning and supply-chain interaction," 2004).
- *Restoration and remanufacturing:* Product remanufacturing and refurbishment are of much significance in circular supply chains as they have an impact on product lifetime. This process usually involves taking apart, cleaning out thoroughly, repairing using small care, reassembling accurately all components of items, and making them look as new as possible. It could often include alterations to comply with new regulations. Restoring users' satisfaction in their ways regarding refurbishment implies systematic ways of cleaning, repairing, and updating them to extend their lifespan, minimize waste generation, and conserve resource base at the same time (Fofou et al., 2021).
- *Recycling:* This is a major component essential for any circular economy that would enable reusing waste resources. Valuable substances such as metals, glass, and polymers are extracted from waste materials during material recovery processes. The economic value of waste materials is maintained by these recycling processes through thermal, chemical, and mechanical reprocessing to make new products or raw materials. It helps improve efficiency at the stage of recycling if those products are designed to allow for recycling and made of components that can easily be separated and reprocessed (Guo et al., 2023).

7.6 FUZZY LOGIC: PRINCIPLES AND APPLICATION

Fuzzy logic is based on the principles of fuzzy set theory and is a complex paradigm in computing. It is used in situations with uncertainty and where there is no clarity, thus supporting management as well as decision-making; it was this theory that led to a huge amount of development related to it (Wu & Xu, 2021).

7.6.1 Fuzzy Logic Principles

Instead of simply black and white, fuzzy logic gives a way to stretch classic logic by truth as a spectrum.

Here are how a few of the concepts have been defined:

- *Fuzzy sets:* A concept "stolen" from math that enables one to represent items that are not specifically identified but belong together in the same group. The idea of a fuzzy set is determined by its membership function $\mu_A(x)$, with x denoting an object, while $\mu_A(x)$ indicates how much object x indeed belongs to a set A (Bustince et al., 2016). The function varies between 0 and 1 showing to what extent an object x belongs to the set:

$$\mu_A : X \rightarrow [0,1]$$

For example, in a fuzzy set that represents a "young age" concept, a person aged 25 would have a membership level equal to .7, indicating some sort of belonging.
- *Membership functions:* The following are the typical examples of membership functions:
Triangular membership function: It is defined as follows:

$$\mu_A(x) = \begin{cases} 0 & \text{if } x \leq a \text{ or } a \geq c \\ \dfrac{x-a}{b-a} & \text{if } a < x < c \\ \dfrac{c-x}{x-b} & \text{if } b \leq x < c \end{cases}$$

where *a*, *b*, and *c* are parameters of the triangle (Yahia et al., 2012).
  - *Gaussian membership function:* It is defined as follows:

$$\mu_A(x) = \exp\left(-\frac{(x-c)^2}{2\sigma^2}\right)$$

where *c* is the average and σ represents the standard deviation according to Wani et al. (2022).
- *Fuzzy rules:* Fuzzy logic systems make use of a set of "if-then" rules to infer outputs by looking at imprecise inputs (Belyadi & Haghighat, 2021). A rule can be expressed in the following format:
If temperature IS high THEN fan speed IS high
  - *Fuzzy inference:* Fuzzy inference is the process of utilizing fuzzy rules to process fuzzy inputs and generate fuzzy outputs (Cao et al., 2024). The method of inference usually entails three successive steps:
    1. *Fuzzification:* Changing exact inputs into fuzzy values by using membership functions.
    2. *Rule evaluation:* Utilizing fuzzy rules to calculate fuzzy outputs based on fuzzy inputs.
    3. *Aggregation:* Aggregating the fuzzy outputs of all rules to create a unified fuzzy set.

- *Defuzzification:* Transforming the combined fuzzy output into a precise numerical number (Natarajan et al., 2023). Typical techniques include the following:
  - Centroid Method:

$$x^* = \frac{\int_a^b x.\mu_A(x)\,\mathrm{d}x}{\int_a^b \mu_A(x)\,\mathrm{d}x}$$

where x^* is the centroid and $[a,b]$ is the range of x.

7.7 APPLICATIONS OF FUZZY LOGIC IN SUPPLY CHAIN MANAGEMENT

According to Ganga and Carpinetti (2011), the use of fuzzy logic in supply chain management has a lot of advantages. Some key benefits are as follows:

- *Demand forecasting:* Fuzzy logic makes it possible to include uncertain and ambiguous information in forecasting models (Salais-Fierro et al., 2020). Furthermore, demand forecast can be expressed using the following formula:

  Forecast (D) = FuzzyInference(Fuzzified_Input _Data)

  In this case, Fuzzy Inference runs acquiescent rules for predicting future demand by using previous information alongside current market trends.
- *Inventory control:* On stock reordering cycle points or order quantities, fuzzy logic can be used to maintain equilibrium between these two (Phruksaphanrat, 2012). This is mathematically expressed as follows:

  Order Quantity (Q) = FuzzyControl(Stock_Level,Demand_Forecast)

  The FuzzyControl changes the number of orders according to the inventory size and changes in demand.
- *Supplier selection:* Some important factors when evaluating suppliers include cost, quality, and time of delivery (Güneri & Deveci, 2023). Such data are transformed into a composite score through fuzzy logic as follows:

  Supplier Score (S) = FuzzyAggregation(Cost,Quality,Delivery_Time)

  In this context, FuzzyAggregation is accountable for merging multiple criteria so that the selection is done effectively.
- *Production planning:* A fuzzy logic–based approach helps in planning efficiently while taking into account constraints like machine availability, human resource input, and market demand (Taghizadeh et al., 2015). An optimal production schedule can be formulated as follows:

  Production Schedule (S) = FuzzyOptimization(Machine_Availability,-Manpower,Demand)

In production planning, there is a need to optimize raw material distribution margins but FuzzyOptimization considers uncertainty surrounding production quantities.

- *Logistics and distribution:* By considering different circumstances, fuzzy logic enhances route allocation accuracy in the logistics sector, hence improving the delivery process in the supply chain (Karabacak & Kutlu, 2024). The best route is formulated thus:

  Optimal Route (R) = FuzzyRouting(Distance,Traffic,Delivery_Urgency)

In evaluating routes based on urgency and traffic distance, FuzzyRouting uses certain fuzzy rules.

7.8 ADVANTAGES OF FUZZY LOGIC IN CIRCULAR SUPPLY CHAINS

In the management of circular supply chains, the use of fuzzy logic offers several benefits:

- *Flexibility:* With the addition of uncertainty and imprecision in supply chain data, it can accommodate complex and changing situations.
- *Robust decision-making:* By combining different elements and uncertainties, fuzzy logic strengthens decision robustness.
- *Enhanced optimization:* By improving production planning, inventory management, and forecasting, it results in decreased wastage and cost-cutting.
- *Enhanced sustainability:* Fuzzy logic is used to promote resource efficiency as a key aspect of the circular economy through closed-loop designs that minimize waste.

7.9 STRATEGIES FOR SUPPLY CHAIN OPTIMIZATION IN CIRCULAR ECONOMY

This section considers strategies that can be used to maximize the efficiency of supply chains within a circular economy. It mainly reflects on methods such as product design for circularity, resource efficiency optimization, closed-loop supply chain implementation, and cooperation between stakeholders (Groenewald, 2024).

The product and process designing methods for circularity involve developing products that can last long, are easily repairable as they were when they were created. On developed strategies, there are so many paths or sets:

- *Modularity:* Things that have parts that can be taken apart easily to repair or upgrade them, thereby increasing both their lifespan and reducing litter.
- *Eco-Design:* Applying design concepts that reduce environmental harm throughout the whole lifespan of the product, such as employing energy-efficient manufacturing techniques and utilizing sustainable materials.

- *Product Life Extension:* Creating items with a focus on easy disassembly, maintenance, and refurbishing, to prolong their lifespan and minimize the demand for new resources.

Optimizing circular supply networks relies heavily on efficient resource utilization. Essential techniques comprise the following:

- *Resource recovery:* Developing procedures to recover valuable resources from discarded products, including metals, polymers, and rare earth elements, to reuse them in the production of new items.
- *Energy efficiency:* Minimizing energy usage in manufacturing and transportation operations by implementing cutting-edge technologies and utilizing sustainable energy sources.
- *Waste minimization:* Implementing strategies to reduce the production of waste, such as lean manufacturing, just-in-time inventory management, and converting trash into energy.

Closed-loop supply chains prioritize the establishment of systems that enable the continuous flow of resources, to reduce waste and optimize resource utilization. Crucial elements comprise the following:

- *Reverse logistics:* Designing streamlined processes to gather and handle returned items for recycling, refurbishing, or remanufacturing.
- *Circular business models:* Implementing business models that encourage circularity, such as product-as-a-service, in which customers lease things rather than buy them, ensuring that they are subsequently returned for reuse or recycling.
- *Life cycle assessment:* Involves detailed evaluations of product life cycles to identify possible resource efficiencies that can be exploited as well as minimizing wastage.

Effective collaboration and coordination among stakeholders play a critical role in achieving successful supply chain optimization in a circular economy (Chauhan et al., 2022). These include the following strategies:

- *Stakeholder engagement:* This involves engaging suppliers, customers, and regulators among other stakeholders about the development and implementation process regarding circular supply chain practices.
- *Cross-sector collaboration:* Organizations across various industries collaborate by sharing knowledge, resources, and best practices for effective circularity.
- *Innovation networks:* Such networks are usually made up of innovators together with scientists. Their main aim is to develop outstanding technologies as well as methodologies meant to optimize circular supply chains.

7.10 CHALLENGES AND SOLUTIONS

Implementing supply chain optimization in a circular economy is difficult largely due to various constraints, including technological, economic, and regulatory issues. These challenges could however be addressed using fuzzy logic which stands as an all-around solution for all these problems. The introduction of new technologies into circular supply chains requires heavy investments and expertise which is not easy to come across in many organizations. On the other hand, it usually needs large initial costs together with the adoption of fresh business models, hence its economic justifiability poses challenges to firms. Circular practices can only be encouraged through appropriate legislation and market conditions but progression is hampered by the absence of clear rules and incentives.

Fuzzy logic provides flexible solutions in dealing with dynamic problems within circular supply chains while at the same time enabling decision-makers to manage inherent unpredictability and intricacy. Additionally, these criteria can be modeled using fuzzy because they can cater for various stakeholders' needs when assessing costs relative to benefits that encompass both economic, environmental, and social angles, hence enabling more rigorous cost-benefit analysis. Through the use of these models, policymakers can determine how best they may regulate their economies to promote a circular economy where waste materials circulate freely. The way forward is to develop mechanisms for mapping out possible effects different policies might have on an economy before adopting them and techniques that can be used to reach these effects.

7.11 CASE STUDIES AND APPLICATIONS

This stage provides extensive instances in which companies and sectors have applied in practice supply chain optimization and logistics in a circular economy with the help of fuzzy logic. These cases serve as a practical illustration of the real uses, results, and benefits derived from these pioneering methods.

CASE STUDY 1: ELECTRONICS INDUSTRY—SUSTAINABLE INVENTORY MANAGEMENT

Efficient electronics industry inventory management entails keeping up with fast technological innovations accompanied by unpredictable demand patterns. In this respect, one of the leading electronics firms uses a fuzzy logic–supported inventory management system to adequately handle unsteady demand and supply situations.

- *Background*: The company experienced difficulties associated with accurate forecasting of demand for electronic components leading to either stock out or overstock.
- *Fuzzy logic application*: By using a fuzzy logic model, the company can adequately treat the uncertainty during sales estimation. The algorithm considered many factors such as historical sales data among other factors like market trends as well as specialists' judgments so that an optimal level could be reached for each particular stage in this process chain.
- *Results*: As a result, there was an inventory holding cost reduction by 20% and also a drop in stockouts by 15% which means that customer satisfaction goes high while operations become efficient.

CASE STUDY 2: AUTOMOTIVE INDUSTRY—WASTE MANAGEMENT OPTIMIZATION

One of the automotive manufacturers was interested in fuzzy logic to improve its waste management strategies, especially on recycling and remanufacturing fronts.

- *Background*: Significant amounts of waste were generated from both production processes and scrapping old cars which compelled the company to improve its waste management practices for environmental reasons as well as economic considerations.
- *Fuzzy logic application*: This process entailed designing a fuzzy decision support model that compares different waste management alternative systems based on parameters like feasibility, cost implication, regulatory compliance, and environmental impact among others.
- *Results*: Consequently, the approach was linked to an improvement in recycling and remanufacturing resulting in a 25% reduction of waste generated in landfills, while increasing raw material usage efficiency by 30%. This action not only reduced the environmental burden but also saved on materials acquisition costs.

7.12 FUTURE DIRECTIONS

The future in the circular economy of supply chain optimization, logistics, and fuzzy logic is influenced by current trends, innovations, and research opportunities. Several forces affect the formation of circular supply networks. The rising adoption of digital technologies like the Internet of Things (IoT), artificial intelligence (AI), and blockchain is transforming supply chain management leading to better visibility, tracking and efficacy. Emerging business models such as product-as-a-service or circular procurement have spearheaded the reuse and recycle cause for items and materials. The use of sustainable technologies and practices sustains the adoption of principles for circular economy across industries (Vimal et al., 2023). There is an expectation that advances in supply chain optimization as well as fuzzy logic will complement circular economy methods in the future. It involves the use of sophisticated data analysis techniques and artificial intelligence for accurate demand forecasting stock control decisions under situations involving uncertainties. In addition, the introduction of smart manufacturing systems, among them, robotics and additive manufacturing, has made it possible for industries to make production processes more flexible, enabling them to achieve higher outputs than before. Other measures to be taken include developing environment-friendly materials alongside improving recycling processes as a way of promoting circular supply chain adoption. Further research is needed to integrate fuzzy logic alongside integrating circular economic concepts into supply chain optimization. Research studies may explore new applications of fuzzy logic in various areas of supply chain management such as risk assessment, supplier selection sustainability appraisal, and so forth. Fuzzy logic integrated with other methodologies like machine learning and optimization techniques can be applied to solving complex supply chain problem sets. The examination of policymaking as well as enactment influences on promotion of circular supply chains through the identification and dissemination of successful approaches for promoting sustainability within these chains (Sánchez-García et al., 2024).

7.13 CONCLUSION

This chapter has looked at how you can take supply chain optimization, logistics, and fuzzy logic and put them under one framework that touches on the concept of a circular economy. It was shown how in a case study, fuzzy logic helps make better decisions when decision-making is complicated, and uncertainty exists. Moreover, the method used here was to pay attention to how best resources could be utilized, waste generated was reduced, while also ensuring a complete use of the system. The actual means and benefits were shown in these examples by showing amazing financial savings made, improved productivity, and reduced environmental impact among others. It is important to keep on researching, innovating, and collaborating to meet the sustainable development goals (SDGs) through the adoption of CE principles. This is evidenced by the way trends are changing and new developments are emerging alongside lots of research opportunities in academia, industry, or even public sector agencies alike; for instance, research on recycling technology—where old can be made new again—or development studies—sustainability through economics rather than environmental science alone. These efforts promote not only a sustainable future for the industry but also have implications for governments and academia.

REFERENCES

Adzija, Robert, and Michael Kukhta. "Chapter 1—The role of logistics in supply chains." Pressbooks, February 28, 2022. https://ecampusontario.pressbooks.pub/logistics001oerfc/chapter/chapter-1/

Ball, Patrick. "Circular economy benefits—Dassault Systèmes blog," July 1, 2024. https://blog.3ds.com/topics/sustainability/10-benefits-of-a-circular-economy/

Bansal, Sanchita, Mansi Jain, Isha Garg, and Mrinalini Srivastava. "Attaining circular economy through business sustainability approach: An integrative review and research agenda." *Journal of Public Affairs* 22, no. 1 (August 19, 2020). https://doi.org/10.1002/pa.2319

Belyadi, Hoss, and Alireza Haghighat. "Fuzzy logic." Elsevier eBooks, 381–418, 2021. https://doi.org/10.1016/b978-0-12-821929-4.00003-2

Bocken, Nancy M.P., Ingrid De Pauw, Conny Bakker, and Bram Van Der Grinten. "Product design and business model strategies for a circular economy." *Journal of Industrial and Production Engineering* 33, no. 5 (April 26, 2016): 308–20. https://doi.org/10.1080/21681015.2016.1172124

Brown, P., C. Von Daniels, N.M.P. Bocken, and A.R. Balkenende. "A process model for collaboration in circular oriented innovation." *Journal of Cleaner Production* 286 (March 1, 2021): 125499. https://doi.org/10.1016/j.jclepro.2020.125499

Bustince, Humberto, Edurne Barrenechea, Miguel Pagola, Javier Fernandez, Zeshui Xu, Benjamin Bedregal, Javier Montero, Hani Hagras, Francisco Herrera, and Bernard De Baets. "A historical account of types of fuzzy sets and their relationships." *IEEE Transactions on Fuzzy Systems* 24, no. 1 (February 1, 2016): 179–94. https://doi.org/10.1109/tfuzz.2015.2451692

Cao, Jin, Ta Zhou, Shaohua Zhi, Saikit Lam, Ge Ren, Yuanpeng Zhang, Yongqiang Wang, Yanjing Dong, and Jing Cai. "Fuzzy inference system with interpretable fuzzy rules: Advancing explainable artificial intelligence for disease diagnosis—a comprehensive review." *Information Sciences* 662 (March 1, 2024): 120212. https://doi.org/10.1016/j.ins.2024.120212

Chauhan, Chetna, Puneet Kaur, Rakesh Arrawatia, Peter Ractham, and Amandeep Dhir. "Supply chain collaboration and sustainable development goals (SDGs). Teamwork makes achieving SDGs dream work." *Journal of Business Research* 147 (August 1, 2022): 290–307. https://doi.org/10.1016/j.jbusres.2022.03.044

Dissanayake, C. Kalpani. "Fuzzy logic applications in supply chain performance measurement." *ResearchGate*, October 1, 2015. https://www.researchgate.net/publication/301956996_FUZZY_LOGIC_APPLICATIONS_IN_SUPPLY_CHAIN_PERFORMANCE_MEASUREMENT

Elisha, Otekenari David. "Moving beyond take-make-dispose to take-make-use for sustainable economy." *ResearchGate*, June 1, 2020. https://www.researchgate.net/publication/349964391_Moving_Beyond_Take-Make-Dispose_to_Take-Make-Use_for_Sustainable_Economy

"Enterprise planning and supply-chain interaction." Elsevier eBooks, 146–84, 2004. https://doi.org/10.1016/b978-075066272-7/50009-9

European Environment Agency. "The benefits to biodiversity of a strong circular economy," n.d. https://www.eea.europa.eu/publications/the-benefits-to-biodiversity

Farooque, Muhammad, Abraham Zhang, Matthias Thürer, Ting Qu, and Donald Huisingh. "Circular supply chain management: A definition and structured literature review." *Journal of Cleaner Production* 228 (August 1, 2019): 882–900. https://doi.org/10.1016/j.jclepro.2019.04.303

Fofou, Raoul Fonkoua, Zhigang Jiang, and Yan Wang. "A Review on the Lifecycle Strategies Enhancing Remanufacturing." *Applied Sciences* 11, no. 13 (June 25, 2021): 5937. https://doi.org/10.3390/app11135937

Ganga, Gilberto Miller Devós, and Luiz Cesar Ribeiro Carpinetti. "A fuzzy logic approach to supply chain performance management." *International Journal of Production Economics* 134, no. 1 (November 1, 2011): 177–87. https://doi.org/10.1016/j.ijpe.2011.06.011

Geissdoerfer, Martin, Paulo Savaget, Nancy M.P. Bocken, and Erik Jan Hultink. "The circular economy—a new sustainability paradigm?" *Journal of Cleaner Production* 143 (February 1, 2017): 757–68. https://doi.org/10.1016/j.jclepro.2016.12.048

Gholizadeh, Hadi, Hamed Fazlollahtabar, and Mohammad Khalilzadeh. "A robust fuzzy stochastic programming for sustainable procurement and logistics under hybrid uncertainty using big data." *Journal of Cleaner Production* 258 (June 1, 2020): 120640. https://doi.org/10.1016/j.jclepro.2020.120640

Ghosh, R. K., M. Ghosh, D. Roy, B. Gupta, S. Chakraborty, A. Biswas, and M. Basu. "Personalized financial services through NLP and AI-driven innovations in FinTech." In *IGI Global eBooks*, 371–420, 2025. https://doi.org/10.4018/979-8-3693-9944-6.ch019

Groenewald, None Elma Sibonghanoy. "Circular economy strategies in supply chain management: Towards zero waste." *Power System Technology* 48, no. 1 (April 17, 2024): 464–80. https://doi.org/10.52783/pst.291

Güneri, Bartu, and Muhammet Deveci. "Evaluation of supplier selection in the defense industry using q-rung orthopair fuzzy set based EDAS approach." *Expert Systems with Applications* 222 (July 1, 2023): 119846. https://doi.org/10.1016/j.eswa.2023.119846

Guo, Jing, Saleem Ali, and Ming Xu. "Recycling is not enough to make the world a greener place: Prospects for the circular economy." *Green Carbon* 1, no. 2 (December 1, 2023): 150–53. https://doi.org/10.1016/j.greenca.2023.10.006

Jauhari, Wakhid Ahmad. "Sustainable inventory management for a closed-loop supply chain with energy usage, imperfect production, and green investment." *Cleaner Logistics and Supply Chain* 4 (July 1, 2022): 100055. https://doi.org/10.1016/j.clscn.2022.100055

Kalebaila, Nonhlanhla, Mpho Kapari, Luxon Nhamo, and Sylvester Mpandeli. "The circular economy as a catalyst for environmental and human health." CRC Press eBooks, 139–58, 2024. https://doi.org/10.1201/9781003327615-7

Kamble, Sachin S, Angappa Gunasekaran, Harsh Parekh, Venkatesh Mani, Amine Belhadi, and Rohit Sharma. "Digital twin for sustainable manufacturing supply chains: Current trends, future perspectives, and an implementation framework." *Technological Forecasting and Social Change* 176 (March 1, 2022): 121448. https://doi.org/10.1016/j.techfore.2021.121448

Karabacak, Ebubekir, and Hüseyin Ali Kutlu. "Evaluating the efficiencies of logistics centers with fuzzy logic: The case of Turkey." *Sustainability* 16, no. 1 (January 4, 2024): 438. https://doi.org/10.3390/su16010438

King, Andrew M., Stuart C. Burgess, Winnie Ijomah, and Chris A. McMahon. "Reducing waste: Repair, recondition, remanufacture or recycle?" *Sustainable Development* 14, no. 4 (December 29, 2005): 257–67. https://doi.org/10.1002/sd.271

Natarajan, Ezhilarasan, Felix Augustin, Mohammed K.A. Kaabar, Charles Robert Kenneth, and Kins Yenoke. "Various defuzzification and ranking techniques for the heptagonal fuzzy number to prioritize the vulnerable countries of stroke disease." *Results in Control and Optimization* 12 (September 1, 2023): 100248. https://doi.org/10.1016/j.rico.2023.100248

Phruksaphanrat, B. "Fuzzy inventory control system for uncertain demand and supply," January 1, 2012. http://www.iaeng.org/publication/IMECS2012/IMECS2012_pp1224-1229.pdf

Rejeb, Abderahman, Andrea Appolloni, Karim Rejeb, Horst Treiblmaier, Mohammad Iranmanesh, and John G. Keogh. "The role of blockchain technology in the transition toward the circular economy: Findings from a systematic literature review." *Resources, Conservation & Recycling Advances* 17 (May 1, 2023): 200126. https://doi.org/10.1016/j.rcradv.2022.200126

Salais-Fierro, Tomas Eloy, Jania Astrid Saucedo-Martinez, Roman Rodriguez-Aguilar, and Jose Manuel Vela-Haro. "Demand prediction using a soft-computing approach: A case study of automotive industry." *Applied Sciences* 10, no. 3 (January 24, 2020): 829. https://doi.org/10.3390/app10030829

Sánchez-García, Eduardo, Javier Martínez-Falcó, Bartolomé Marco-Lajara, and Encarnación Manresa-Marhuenda. "Revolutionizing the circular economy through new technologies: A new era of sustainable progress." *Environmental Technology & Innovation* 33 (February 1, 2024): 103509. https://doi.org/10.1016/j.eti.2023.103509

Taghizadeh, Houshang, Ardeshir Bazrkar, and Mohammad Abedzadeh. "Optimization production planning using fuzzy goal programming techniques." *Modern Applied Science* 9, no. 9 (July 8, 2015). https://doi.org/10.5539/mas.v9n9p68

Takacs, Fabian, Dunia Brunner, and Karolin Frankenberger. "Barriers to a circular economy in small- and medium-sized enterprises and their integration in a sustainable strategic management framework." *Journal of Cleaner Production* 362 (August 1, 2022): 132227. https://doi.org/10.1016/j.jclepro.2022.132227

Velenturf, Anne P.M., and Phil Purnell. "Principles for a sustainable circular economy." *Sustainable Production and Consumption* 27 (July 1, 2021): 1437–57. https://doi.org/10.1016/j.spc.2021.02.018

Vimal K.E.K., K. Sivakumar, Jayakrishna Kandasamy, Vaibhav Venkat, and Raghuraman Srinivasan Mani. "Barriers to the adoption of digital technologies in a functional circular economy network." *Operations Management Research/Operations Management Research: Advancing Practice Through Research* 16, no. 3 (April 25, 2023): 1541–61. https://doi.org/10.1007/s12063-023-00375-y

Wani, Zubair Rashid, Manzoor Tantray, Ehsan Noroozinejad Farsangi, Nikolaos Nikitas, Mohammad Noori, Bijan Samali, and T.Y. Yang. "A critical review on control strategies for structural vibration control." *Annual Reviews in Control* 54 (January 1, 2022): 103–24. https://doi.org/10.1016/j.arcontrol.2022.09.002

Wu, Hangyao, and Zeshui Xu. "Fuzzy logic in decision support: Methods, applications and future trends," 2021. https://univagora.ro/jour/index.php/ijccc/article/view/4044

Yadav, Sudhir Kumar Singh. "Supply chain management: A tool of business process integration." *ResearchGate*, November 1, 2013. https://www.researchgate.net/publication/299777734_Supply_Chain_Management_A_Tool_of_Business_Process_Integration

Yahia, Nesrine Ben, Narjès Bellamine Ben Saoud, and Henda Ben Ghezala. "Integrating fuzzy case-based reasoning and particle swarm optimization to support decision making." *ResearchGate*, May 1, 2012. https://www.researchgate.net/publication/265798833_Integrating_fuzzy_case-based_reasoning_and_particle_swarm_optimization_to_support_decision_making

Yang, Mingyu, Lin Chen, Jiangjiang Wang, Goodluck Msigwa, Ahmed I. Osman, Samer Fawzy, David W. Rooney, and Pow-Seng Yap. "Circular economy strategies for combating climate change and other environmental issues." *Environmental Chemistry Letters* 21, no. 1 (September 6, 2022): 55–80. https://doi.org/10.1007/s10311-022-01499-6

8 Circular Product Design and Development
Principles, Strategies, and Case Studies

Tran Minh Tung

8.1 INTRODUCTION TO CIRCULAR ECONOMY AND ITS IMPORTANCE IN PRODUCT DESIGN

In a linear economy, products are designed to be worn out or replaced and are discarded after a short period of use, resulting in social, environmental, and economic challenges. However, a circular economy approach proposes system solutions with the promotion of product longevity and reutilization as well as remanufacturing. Although the transition from a linear economy to a circular economy is not a simple task and there is little experience disseminated, there are great opportunities for both the industry and consumers. Considering that a key element in the exploitation of these opportunities is based on product design, this chapter provides an overview of some principles, concepts, benefits, and challenges of designing products for a circular economy, as well as some exemplary case studies (Reslan et al., 2022) (Neves and Marques, 2022).

The circular economy enhances the capacity of natural systems to regenerate and promote system effectiveness with the slow-loop cycles of the reuse and repair functions, so that the products are used to their maximum value, ultimately transitioning into more linear singular application. Despite being in its early developmental stages, the necessity of forming its theoretical purpose and other concept-accepting points of view on a widely accepted scale highlights the core principles of circular products. Products should be designed and used in ways that make them as long-lasting as possible, thereby increasing their initial investment value. In practical instruction, circular economy and design-for-philosophy generally concentrate on the extension of product life as well as resulting in the regeneration and reuse of components and materials. Generally, whether reducing usage of natural resources, enhancing resource productivity, as well as increasing the social sector, the positive effects of the circular economy are categorized as technical, ecological, economic, and social in many cases (Velenturf and Purnell, 2021; Stefanakis et al., 2021).

DOI: 10.1201/9781003538837-8

8.1.1 Definition and Principles of Circular Economy

The transition to a more sustainable society urges the shift from the current linear economic system to a circular one, allowing production systems and societies to become more sustainable. There are many definitions of the circular economy (CE), however, a commonly accepted one is given by the Ellen MacArthur Foundation. Here, the circular economy refers to an economic system that "aims to keep products, components, and materials at their highest utility and value at all times." Brekke and Whalen outlined the fundamental principles of the circular economy, emphasizing the value chain and the aid to develop more sustainable products and services (Brekke, 2021; Whalen and Whalen, 2020).

The CE concept triggers many principles regarding the way product design and development should be realized. The first and foremost are those of product longevity and component life extension that provide for the preservation of both technical and economic value inside products. Furthermore, the principles of diversified use strategies trigger the development of multifunctional materials and components. One of the consequences is the necessity for product disintegrity design, that is, design elements that would enable the disintegration of large systems into smaller and more standardized modules. Decentralized systems are more flexible and adaptable to diversified user needs and infrastructural conditions. Moqri stated that the development of mass customization and products adapted to local production lines is still in its infancy; however, the guiding principles of this radical change in product development are given in the definition and principles of the circular economy (Moqri et al., 2023).

8.1.2 Benefits and Challenges of Implementing Circular Economy in Product Design

Considering the pace of consumption, a circular economy design would help in ensuring long-term sustainability of resources. Manufacturing products, which last longer, are more easily upgradeable and repairable, can be returned and refurbished or remanufactured, and then recycled at their end-of-life, would result in the use of fewer natural resources. Circular economy is a new system, and new systems bring both new opportunities and challenges with them. While the end goal of the circular is to be resource-efficient, the path to achieve that goal is different from what has been done until now. It involves moving beyond traditional regulatory-based control and policy action, and engaging in fostering new economic activities, products, services, and new business models, which could be a daunting task (Dey et al., 2022).

The detailed listing of challenges faced in the design of circular products is presented in Marsh et al. (2022). The enhanced interest in the topic in recent years has led to increased attention by researchers and practitioners in identifying ways to overcome these challenges. Although circular design might be an important solution for reducing the environmental impact from products, the following points have to be taken into account. Integration of circular economy into product design is still at its beginning. No methodology for a specific method for connection between the product design and end-of-life is defined. Economic benefits for companies have to be proved yet. Costs for treatment of a product, or at least information about the treatment, have to be available when designing the product.

8.2 FUNDAMENTAL CONCEPTS IN PRODUCT DESIGN AND DEVELOPMENT

Typically, a formal definition of the "product" concept does not exist in the literature; however, available definitions differ from definition to definition. Items having finished goods such as consumer goods, business goods, and durable goods, which fit the concept of a product, are given as canned, controllable, and switchable where the services component is dominant. Service systems can be assessed from the three main components: complementary product values, fundamental values, and services. Products are provided to customers in return for other products or services such as the sales of manufacturing or engineering service operations and cooperation parallel to the product (Jaakkola, 2020).

A product is a combination of goods and services. Products that provide the desired features are called goods, and the outputs are called services. The product is not only in the form of material and power but can also be in the form of a tower. New products in companies are only a success with good organization of a series of activities to obtain the desired production. New product development consists of a set of stages parallel to the existing activities of the company. This information is pre-information to understand and internalize the continuing processes. Companies supply products to their customers in exchange for the money required. A business's most basic characteristic is to put products produced as goods or services in the market, provide them to customers, and carry out all of this profitably. Companies make a variety of products available to customers who differ among themselves and want different characteristics in the targeted and expected use (Jeanrenaud and Romanazzi, 2024; Chen et al., 2024).

8.2.1 Life Cycle Assessment

Life cycle assessment (LCA) is one of the crucial aspects for designing and developing sustainable and circular products. It is also a pivotal method to develop eco-innovation which provides help to evaluate the environmental impact of a product or service throughout its life cycle. LCA is an extended method from ecological cost accounting: ECA is the most valuable tool to implement the principles of green manufacturing in the industrial sectors. It also provides investigation effects of alternative final treatment options considering batteries in several different categories of adverse air emissions, waterborne effluents, and residual waste, energy consumption, and environmental impacts (Pacana et al., 2023).

Analogous to the ISO 9000 quality standards, a minimum requirement for LCA would contain ISO 41041, which has also been approved by EUROLCA and has been formal statement, "Part 1: LCA standard chains simplified and straightforward checklist and questions and "Part 2: Examples," including Transportation, Passenger Car Pilots, Automobiles, Electronics, Waste Management, Intertwined Green Energy Consumption, Design, Crop Protection, Newspapers, Household Products, Real Estate, Food, Garment, and Shelter. It contains recommendations due to the use of Table 8.1 for developing certain LCA studies, and this protocol would also be generally applicable. It additionally refers to the general literature and recommendations that would provide resources and guidance for the development of LCA in a fashion (Gupta, 2023).

TABLE 8.1
Summary of Key Literature on Circular Product Design and Development

No.	Authors	Topic	Main Finding
1	Reslan et al., 2022	Circular Economy and Its Importance in Product Design	Circular economy addresses social, environmental, and economic challenges of linear economy. Product design is crucial for exploiting opportunities in circular economy
2	Neves and Marques, 2022	Circular Economy and Its Importance in Product Design	Circular economy emphasizes product longevity, reuse, and remanufacturing. Transition to circular economy is challenging but offers significant opportunities
3	Velenturf and Purnell, 2021	Circular Economy and its Importance in Product Design	Circular economy enhances natural regeneration and system effectiveness. Product design should focus on longevity and component reuse.
4	Stefanakis et al., 2021	Circular Economy and Its Importance in Product Design	Circular economy benefits include reduced resource use, increased resource productivity, and social improvement
5	Brekke, 2021	Definition and Principles of Circular Economy	The circular economy is an economic system that aims to keep products, components, and materials at their highest utility and value at all times
6	Whalen and Whalen, 2020	Definition and Principles of Circular Economy	Circular economy emphasizes the value chain and the aid to develop more sustainable products and services
7	Moqri et al., 2023	Definition and Principles of Circular Economy	Circular economy principles include product longevity, component life extension, diversified use strategies, product disintegrity design, and decentralized systems
8	Dey et al., 2022	Benefits and Challenges of Implementing Circular Economy in Product Design	Circular economy design helps ensure long-term sustainability of resources. Challenges include moving beyond traditional regulatory-based control and policy action
9	Marsh et al., 2022	Benefits and Challenges of Implementing Circular Economy in Product Design	Challenges of circular design include the lack of a specific methodology for connecting product design and end-of-life, unproven economic benefits for companies, and unavailable information about product treatment costs
10	Jaakkola, 2020	Fundamental Concepts in Product Design and Development	A product is a combination of goods and services. Products that provide the desired features are called goods, and the outputs are called services

11	Jeanrenaud and Romanazzi, 2024	Fundamental Concepts in Product Design and Development	New products in companies are only a success with good organization of a series of activities to obtain the desired production
12	Chen et al., 2024	Fundamental Concepts in Product Design and Development	Companies supply products to their customers in exchange for the money required. A business's most basic characteristic is to put products produced as goods or services in the market, provide them to customers, and carry out all of this profitably
13	Pacana et al., 2023	Life Cycle Assessment (LCA)	LCA is a crucial aspect for designing and developing sustainable and circular products. It is also a pivotal method to develop eco-innovation which provides help to evaluate the environmental impact of a product or service throughout its life cycle
14	Gupta, 2023	Life Cycle Assessment (LCA)	A minimum requirement for LCA would contain ISO 41041, which has also been approved by EUROLCA's approval and has been formal statement, "Part 1: LCA standard chains simplified and straightforward checklist and questions and "Part 2: Examples," including transportation, passenger car pilots, automobiles, electronics, waste management, intertwined green energy consumption, design, crop protection, newspapers, household products, real estate, food, garment, and shelter
15	Bai et al., 2024	Design Strategies for Circular Products	Several interesting design strategies to enable the development of a circular product, which follows the principles of a circular economy, have been envisaged, such as the design and development of products taking full account of disassembly possibility; disassembly by separation methods, in particular dry detaching technologies or compatibilization strategies are suggested; the use of natural and recycled materials, and/or formulation of products and/or reinstating of such a strategy based on TRL, based on digital twin tools; the design of fostering modularity both from a physical point of view, by means of modular design in products, animate or tools used for manufacture and disassembly, and from artificial modular design software; the new trend of servitization, namely, the design of Circular Service Systems, that enable the reduction for the product used

(Continued)

Table 8.1 (*Continued*)
Summary of Key Literature on Circular Product Design and Development

No.	Authors	Topic	Main Finding
16	Eberhardt et al., 2022	Design Strategies for Circular Products	The novelty of the approaches for enabling circular product design is based on the repositioning of the EoL through the use of an artificial intelligence (AI) enabler that aims at take-back, reassembling, and distributing back to the point of reuse, all over the world
17	Ostapska et al., 2024	Design for Disassembly and Reassembly	Disassembly refers to the separation of two or more assembled products from each other. Each disassembled part can be directly recycled, repaired, or rebuilt. Design for disassembly, in turn, refers to the design of products so that they can be easily disassembled for maintenance and repair
18	Ullah et al., 2021	Design for Disassembly and Reassembly	Disassembly mainly occurs in the reverse supply chain, which includes receive products; disassemble, sort, and recycle components; reassemble components into new products; assembly and remanufacturing refer to the green supply chain management and circular economy
19	Burke et al., 2023	Design for Durability and Longevity	Durability refers to the ability of a product to remain in good working order for a long period of time, whereas longevity refers to the overall lifespan of a product.
20	Xie et al., 2024	Design for Durability and Longevity	Design for durability and longevity can be realized through various strategies. These include selecting reliable materials/components with long life cycles, reinforcing "hot spots" of products to avoid potential malfunctioning, designing mechanical components with redundancy to maintain the functionalities of products, preventing misuse to extend the lifespan of products, and creating an emotional bond with consumers to improve a product's value perception and thus the preservation, care, upgrading, repair, and reuse
21	Praveena et al., 2022	Material Selection and Sourcing for Circular Products	Conventionally, cost, technical requirements, and performance have long determined material selection in the design discipline. Based on conventional criteria, a design team considers better options over and above cost-effectiveness for incorporating recycled and reuse materials in product designs

22	Raj et al., 2022	Material Selection and Sourcing for Circular Products	In terms of environmental performance, recycled and renewable content products are known to be valuable for their extended environmental benefits and circularity in particular. Thus, sustainable and circular goals are being communicated increasingly in material selection and sourcing guidelines accompanying product design and circular product design literature
23	Rai et al., 2021	Material Selection and Sourcing for Circular Products	The excessive use of virgin resources should be averted as indicated in circularity principles and guidelines. In accordance with that principle, the use of recycled and renewable materials is voiced as primarily mandatory in design principles accompanying circularity principles
24	Forastero, 2023	Material Selection and Sourcing for Circular Products	The aforementioned principles seem to support a "lightened" approach to material selection and sourcing since it is the extra use of material that is a recycled or renewable resource that could actively support a circular economy ideally made possible through design. In contrast, the inferiority or underperformance of these materials could pass under the radar when considering these materials and how to include them in designing for circularity
25	Lamba et al., 2022	Recycled and Renewable Materials	The utilization of recycled materials from end-of-life products, industrial waste, or post-consumer waste can not only help to reduce the cost of products, but also contribute to the durability, sustainability, and circularity
26	Khorasani et al., 2022	Manufacturing Processes and Technologies for Circular Products	At the intersection of technology and business, additive manufacturing is highly influential in shaping new business, management, and innovation opportunities. Together with a shift toward more sustainable, customized, and efficient solutions, 3D printing is becoming one of the most prominent strategies and technologies in the field of circular product design and development
27	Hegab et al., 2023	Manufacturing	

8.3 DESIGN STRATEGIES FOR CIRCULAR PRODUCTS

Even if there is no consensus on a global methodology for the circularity strategies in products, several interesting design strategies to enable the development of a circular product, which follows the principles of a circular economy, have been envisaged: the design and development of products taking full account of disassembly possibility; disassembly by separation methods, in particular dry detaching technologies or compatibilization strategies; the use of natural and recycled materials, and/or formulation of products and/or reinstating of such a strategy based on TRL, based on digital twin tools; the design of fostering modularity both from a physical point of view, by means of modular design in products, animate or tools used for manufacture and disassembly, and from artificial modular design software; the new trend of servitization, namely, the design of Circular Service Systems, that enable the reduction for the product used (Bai et al., 2024).

The novelty of the approaches for enabling circular product design is based on the repositioning of the EoL through the use of an artificial intelligence (AI) enabler that aims at take-back, reassembling, and distributing back to the point of reuse, all over the world. The focus is on the development of design approaches that are able to inform the designer at the early phases, taking into account that decisions made in these phases will have a major impact on the competitiveness of a company as, on average, about 80% of manufacturing costs are already committed in early design phases. This would provide designers with an alternative to a largely unsuccessful and incomplete range of methods. The general objective of such a strategy consists in developing a business model that makes remanufacturing possible for green energy devices with the following European characteristics: high quality, as they are still working; expensive, high level of technology that makes complex to recover the critical raw material in the junkyard; guide design to highlight the effectiveness of the strategies. By referring to the case studies above, we can conclude that efficient waste management relies heavily on a product's conceptual design. A variety of studies have highlighted the fact that only by changing the way product design is conceived can a product's end of life be managed effectively. In creating a product that is recyclable, part of a reuse loop or composed of materials that can be remanufactured, the product will be considered more valuable. Thus, it is the conceptual designs that are needed to produce such a product, and this calls for them to be made in the early stages of a product's life cycle. Strategic design seems to be a good choice in designing products that can be addressed in the early stages of design (Eberhardt et al., 2022).

8.3.1 Design for Disassembly and Reassembly

Disassembly refers to the separation of two or more assembled products from each other. Each disassembled part can be directly recycled, repaired, or rebuilt. Design for disassembly, in turn, refers to the design of products so that they can be easily disassembled for maintenance and repair. It is good for product availability, serviceability, and reduces cost and purchased part repeatability. Also, it will be helpful in the end of use stage, where the products can be disassembled to recover materials and for component reuse. The results will impact the objectives of waste reduction and cost-effective strategies. The impacts of disassembly on maintenance, repair, and ease of recycling are typical life cycle characteristics. They complete product capacity to regenerate the useful and valuable essence, these normal or unintegrated information (Ostapska et al., 2024).

Disassembly mainly occurs in the reverse supply chain, which includes receive products; disassemble, sort, and recycle components; reassemble components into new products; assembly and remanufacturing, refer to the green supply chain management and circular economy. Besides, the reassembly process is performed not only in the remanufacturing step but also while designing a new product by incorporating an existing component or a module, then the reassembly process also starts when the product leaves the factory floor after reuse recovery such as upgrade, remanufacturing, or the end of use. Reassembly can have direct economic advantages, including the reduction of inventory costs and the exploitation of cost-efficient product variants. Moreover, it is an alternative end-of-life strategy that promotes sustainability and circularity. In this case, different solutions are possible, and their relevance is significantly influenced by the environmental and economic conditions (Ullah et al., 2021).

8.3.2 Design for Durability and Longevity

Durability refers to the ability of a product to remain in good working order for a long period of time, whereas longevity refers to the overall lifespan of a product. Under the principles of circularity and sustainability, longevity aims to reduce the frequency of consumption and minimize the amount of waste production. It is about the quantity. Apart from quantity, durability is about the quality of a product. Durability means that a product is of high quality and performs its original intended functions during its entire lifespan (Burke et al., 2023).

Design for durability and longevity can be realized through various strategies. These include selecting reliable materials/components with long life cycles, reinforcing "hot spots" of products to avoid potential malfunctioning, designing mechanical components with redundancy to maintain the functionalities of products, preventing misuse to extend the lifespan of products, and creating an emotional bond with consumers to improve a product's value perception and thus the preservation, care, upgrading, repair, and reuse (Xie et al., 2024).

In conclusion, rather than being short-lived or easily worn out, durable and long-lasting products require a rigorous combination of design principles, driven by both numbers (intended lifespan, regularity of usage) and emotions of the users (aesthetic and emotional appeal).

8.4 MATERIAL SELECTION AND SOURCING FOR CIRCULAR PRODUCTS

Materials are crucial components of developing products. Conventionally, cost, technical requirements, and performance have long determined material selection in the design discipline. Based on conventional criteria, a design team considers better options over and above cost-effectiveness for incorporating recycled and reused materials in product designs (Praveena et al., 2022).

In terms of environmental performance, recycled and renewable content products are known to be valuable for their extended environmental benefits and circularity in particular. Thus, sustainable and circular goals are being communicated increasingly in material selection and sourcing guidelines accompanying product design and circular product design literature. Inclusion of recyclability and renewability percentages of potential candidate materials, materials with lower embodied primary energy, and negative (eco-)toxicology for crowds, for example, become one of the key criteria to prefer materials over others for product design (Raj et al., 2022).

The excessive use of virgin resources should be averted as indicated in circularity principles and guidelines. In accordance with that principle, the use of recycled and renewable materials is voiced as primarily mandatory in design principles accompanying circularity principles. The use of such materials that come from abnormal places, such as other products, should be actively pursued in order to achieve better circularity in products, as suggested by the American National Standards Institute (ANSI), for example (Rai et al., 2021).

Therefore, the aforementioned principles seem to support a "lightened" approach to material selection and sourcing since it is the extra use of material that is a recycled or renewable resource that could actively support a circular economy ideally made possible through design. In contrast, the inferiority or underperformance of these materials could pass under the radar when considering these materials and how to include them in designing for circularity.

This chapter investigates all three design and innovation processes to gain an understanding of the design community's concentration on materials and what is being communicated about this field in best practices involving design innovation and circular economy for product design.

8.4.1 Recycled and Renewable Materials

The utilization of recycled materials from end-of-life products, industrial waste, or post-consumer waste can not only help to reduce the cost of products, but also contribute to the durability, sustainability, and circularity. The use of recycled materials helps to transform waste into free feedstock for production, saving raw materials and reducing energy consumption and greenhouse gas emissions. Also, recycled materials are often reported to have better mechanical properties than equivalent virgin powders, such as 316L stainless steel, polymers, and ceramics. The use of renewable materials as an alternative to petroleum-based polymers and metals can promote ecological and economic sustainability and contribute to supply resistance through conservation of nonrenewable resources, such as metals and rare earth elements (Lamba et al., 2022).

Criteria and process for the selection of renewable and recycled resources are shown in the standard ASTM D6866–18 and BS 8102, respectively. It is first suggested to select renewable and recycled materials with low embodied energy at the production stage, and next to take into account other product design-related issues, such as mechanical property, life cycle, environmental impacts, and health hazard. In general, industries tend to prefer virgin materials to recycled ones owing to shortage of supply, fluctuation of cost, low purity, and so on. A large number of companies are considering renewable and recycled materials, for example, in the manufacture of cups, construction materials, and fashion. Cotton, wool, and polyester, derived from renewable resources, and man-made cellulose fiber are the major fiber types utilized in leggings, jackets, and sweaters.

8.5 MANUFACTURING PROCESSES AND TECHNOLOGIES FOR CIRCULAR PRODUCTS

Traditionally, manufacturing processes and end-of-life handling strategies influenced product design and development to a large extent. However, when a circular product is the primary goal, we often have to develop and shift a "product-service system" toward more sustainable and resource-connected designs. At the intersection of technology and business, additive manufacturing is highly influential in shaping new business, management, and innovation opportunities. Together with a shift toward more sustainable, customized, and efficient solutions, 3D printing is becoming one of the most prominent strategies and technologies in the field of circular product design and development. This section provides a comprehensive look at additive manufacturing, its materials, and methods of 3D printed recyclability take-back systems and process flow diagram (Khorasani et al., 2022; Hegab et al., 2023).

Additive manufacturing (AM) is the system where a 3D solid object is printed vertically using thin consecutive layers sliced from a 3D computer-aided design model in a computer-aided manufacturing toolpath. 3D printing permits the quick production of a physical mock-up from computer designs in hours, as opposed to days or weeks using more traditional, costly techniques. Additionally, AM deleverages broad tooling requirements, where normally single-use tooling is engaged, thereby creating waste and leading to significant footprint costs. Direct costs include the tooling itself and the labor required to machine them into configurations. Employing AM can allow manufacturers to completely avoid the use of these metals and minimize the resultant footprint. In addition, AM systems can also accommodate design aids into products or components in such a way that would be impossible with conventional manufacturing. For end-user markets, companies can produce fully functioning replicas of niche parts that are no longer in production. AM can tenably connect to the ideals of circularity while offering numerous environmental and cost-effective benefits (Kumar et al., 2023).

8.5.1 Additive Manufacturing (3D Printing)

Three-dimensional printing is one of the emerging and pivotal technologies that can be used in the circular development of products. It can help in various ways in the quest for achieving the principles of circular economy. Some renowned product companies such as Adidas, Marchese, and H&M are already producing products utilizing 3D printing. They are usually personalized and tailored to meet individual customers' needs. This application has triggered the development of a vast array of what are called bespoke, tailor-made, customized products that are inherently sustainable as one product is designed and personalized for a single user. These are not mass-produced, either they could last longer or these products are made in such a way to last longer and used for reuse, refurbishment, remanufacture, repair, and recycling. Three-dimensional printing technology enables us to break free from the constraints associated with traditional manufacturing and offer endless possibilities when it comes to product design, functionality, and personalization (Palladinetti, 2023; Tripathi and Gupta, 2020).

One key factor that contributes to the success of any innovative technology in establishing a competitive edge is its suitability or readiness for the application and exploitation in the meeting of the objectives. In this regard, 3D printing has traits that make it perfectly suitable for the circular product development and production; consequently, its contribution toward the creation of sustainable products and operations. Some key features and characteristics of 3D printing that render it an excellent tool and technology to meet the sustainable business objectives include producing less waste, manufacturing complex, customized, and personalized parts, manufacturing parts with less energy, and providing more functional solutions with fewer parts.

8.6 END-OF-LIFE STRATEGIES AND CIRCULAR ECONOMY MODELS

End-of-life strategies, and thereby strategies for recovering products after use, are crucial for establishing product sustainability throughout the product's life cycle. Two primary end-of-life strategies are take-back systems and reverse logistics. The main objective behind such strategies is recycling, remanufacturing, or refurbishing of products, components, and materials. Already implemented in many industries around the world, take-back systems vary in the sectors they serve, the legal and regulatory arrangements that direct their operation, and specific waste management problems they address. Take-back operations have been linked to a variety of business models and can be considered a component of several circular economy models (Mallick et al., 2023; Butt et al., 2023).

In circular economy models, a closed-loop circulation strategy, known as the maintenance, repair, and recovery, is a central feature. It is, in fact, the end-of-life product and by-products management option, along with product cascading and reuse, under which a product or material can revert to earlier life cycle stages. In addition, the reverse logistics that move products from the end user or customer back to manufacturing or refurbishment locations can be undertaken to support two other key circular economy models or principles: upgrading the rate of materials and component reuse and recycling at lower states of aggregation and achieving higher order waste efficiencies by creating a multidirectional materials recovery process. Products and by-products have clearly established material or embodied energy value. In a circular economy, it is desirable to extract that value in recovered product, component, and material flows and ensure the residual waste or pollution is recognized, usually as one element in holistic industrial ecology solution (Pichlak and Szromek, 2022).

8.6.1 Take-Back Systems and Reverse Logistics

The implementation of a circular economy for products requires the safe recovery of the products which have been out there in society and the economy for many years. However, only when this is already planned will a manufacturer have the motivation and be able to include it in the design specifications. Products come in a wide range of shapes and sizes, and the technologies needed to recover them and reprocess them are also varied. In sum, in a new production-based circular economy, the take-back, reverse logistics, and eventual reprocessing system is essential, and it can be expected that recovery and recycling companies will partner with the product manufacturers to make these operations smoother. Hence, a take-back system is also essential to the overall circular approach, and several design strategies to make it work better will be discussed in the next section (Hagelüken and Goldmann, 2022).

Take-back systems are not always available for disassembly, refurbishing, and reuse—with some exceptions for certain limited industries and some specific products. As noted above, products are rarely designed to be durable, so there are few worth taking back long after the end of a product's useful life. Products, components, and systems often need considerable energy, time, and expense to recover them alone. Take-back of the product is not the first option to deal with at the end of the useful life classification. The take-back of the product will often only occur after the very heavy and complicated set of processes outlined in the take-back diagram mentioned have taken place. Consequently, it would be misleading to talk about "take-back" as an end-of-life option unless all these operations are also considered. Only at the end of this time sequence, when there is less than 5% remaining of the original product mass, are the discussions really getting onto recovery and eventual final disposal rather than take-back (Andersen et al., 2023; Schepler et al., 2024).

8.7 CASE STUDIES IN CIRCULAR PRODUCT DESIGN

Our first example comes from the fashion industry where designers, companies, and producers start to adopt closed-loop fashion development to consume less resources and energy on one hand, and to be able to benefit from rapid technological developments, on the other. As a pioneer textile company in adopting DE, I:CO gives different examples on their website showing how they update fashion products (e.g., T-shirts) before returning them to the stores. First, based on DE, the company began to give "Swiss army socks" made of 35% recycled material, and then extended their portfolio with materials made of 70% and 100% recycled fibers. This is because feedback showed that socks got worn out pretty quickly. T-shirts, another daily fashion element, are produced by following design for modularity principles. Their designs are slightly altered, and then they are put back into stores for purchase. In sum, DE not only increases resource utilization in fashion and textiles, but it also spurs fashion designers to produce modular products to be able to modify them during the use phase (Dragomir and Dumitru, 2022).

ReGenerate and ProQuaDrais are two interdisciplinary research projects that overlap with new opportunities offered to the current and future apparel industry in Switzerland. Both project goals are to help organizations in this sector, including learning networks, turn a series of textiles and fashion concepts and principles into operational procedures, products, and services. ProQuaDrais investigates how to use recycled clothing to create new quality clothing in a unique style, and analyzes in detail how the apparel sector currently works in Switzerland and Europe. ReGenerate developed practical methods and space to generate concepts, products, ideas, and visions for a reuse (right before recycling) phase. It brought together 14 Swiss companies (clothing, textiles, and laundry services) and 7 researchers from two Swiss research institutions from the technical and social science fields.

Case 1: Sustainable Fashion and Textiles—Fostering Circularity through Design and Business Model Innovation: This case study refers to the implementation of circular design principles in the context of the fashion and textile industry. The authors confront the ecological costs and issues with the market for fast fashion, which enables consumers to select and buy from rapidly changing lines of inexpensive wearables. Authors contribute with insights from novel interdisciplinary research in combining perspectives from design and market diffusion to impact the uptake of circular economy principles in product design to enhance market for reuse, repair, and recycling of fashion and textiles (Figure 8.1).

Case 2: Implementing Circular Economy Principles in Design Practice—A Case Study of Three Circular Design Experiments: Stephen Clune, Anna Ekström, Rickard Luckit, and Beatrix Breuss describe experiments conducted with the aim to explore opportunities for the functional and profitable implementation of circular economy (CE) strategies through the use of eco-design and product service system (PSS) development tools. The focus of this work is to describe practical involvement in these experiments and to display real-world examples resulting from the CE design methodology. In this chapter, they have had access to a product developer and a product development company and describe their experiences in using the methodology of implementing circular economy principles into design practice. The subjected designer wishes to explore and develop a diverse collection of shoes that stand

FIGURE 8.1 Infinited fiber wet spinning process. (© Infinited Fiber Company, 2023, "Clothing Manufacturers Aim to Get Fashionable with Greener Practices.")

the test of time. However, the single founder business model dictates an ongoing need for sales in order to sustain the business. Furthermore, the small company (currently, the business is a sole proprietorship, with long-term ambitions for growth) is reliant on the supply chain services of a large-scale production partner (Dey et al., 2022).

8.8 CONCLUSION

The transition from a linear to a circular economy necessitates a fundamental shift in product design and development. This chapter has explored the core principles, strategies, and challenges involved in creating circular products. By examining the critical role of product design in resource efficiency and waste reduction, it becomes evident that a circular approach offers significant potential for environmental and economic benefits.

Key findings highlight the importance of product longevity, modularity, and ease of disassembly in achieving circularity. The incorporation of recycled and renewable materials, coupled with advanced manufacturing techniques like additive manufacturing, presents promising avenues for sustainable product development. Moreover, end-of-life strategies, such as take-back systems and reverse logistics, are essential components of a circular economy.

While challenges such as economic viability and infrastructure limitations persist, the growing body of research and industry adoption demonstrates a clear trajectory toward circular product design. By integrating circular principles throughout the product life cycle, from conception to end-of-life, businesses can contribute to a more sustainable future while gaining a competitive advantage.

Future research should focus on developing standardized methodologies for circular product design, addressing the economic feasibility of implementing circular strategies on a large scale, and exploring innovative business models to incentivize circularity. Ultimately, a collaborative effort involving policymakers, industry, and academia is required to accelerate the transition to a circular economy.

By embracing circular product design, we can move toward a future where resources are utilized efficiently, waste is minimized, and products are designed for continuous value creation.

REFERENCES

Andersen, Ann-Louise, Thomas D. Brunoe, Markus Thomas Bockholt, Alessia Napoleone, Jesper Hemdrup Kristensen, Michele Colli, Brian Vejrum Wæhrens, and Kjeld Nielsen. "Changeable closed-loop manufacturing systems: challenges in product take-back and evaluation of reconfigurable solutions." *International Journal of Production Research* 61, no. 3 (2023): 839–858. https://doi.org/10.1080/00207543.2021.2017504

Bai, Chunguang, Qingyun Zhu, and Joseph Sarkis. "Circular economy and circularity supplier selection: a fuzzy group decision approach." *International Journal of Production Research* 62, no. 7 (2024): 2307–2330. https://doi.org/10.1080/00207543.2022.2037779

Brekke, T. "The circular economy." *Foundations of a Sustainable Economy* (2021). https://www.taylorfrancis.com/chapters/edit/10.4324/9781003010579-4/circular-economy-thomas-brekke

Burke, H., A. Zhang, and J. X. Wang. "Integrating product design and supply chain management for a circular economy." *Production Planning & Control* (2023). https://doi.org/10.1080/09537287.2021.1983063

Butt, A. S., I. Ali, and K. Govindan. "The role of reverse logistics in a circular economy for achieving sustainable development goals: A multiple case study of retail firms." *Production Planning & Control* (2023). https://doi.org/10.1080/09537287.2023.2197851

Chen, Jin, Zheng Liu, Xu Huang, Chenwang Wu, Qi Liu, Gangwei Jiang, Yuanhao Pu et al. "When large language models meet personalization: Perspectives of challenges and opportunities." *World Wide Web* 27, no. 4 (2024): 42. https://doi.org/10.1007/s11280-024-01276-1

Dey, Prasanta Kumar, Chrisovaladis Malesios, Debashree De, Pawan Budhwar, Soumyadeb Chowdhury, and Walid Cheffi. "Circular economy to enhance sustainability of small and medium sized enterprises." In *Supply chain sustainability in small and medium sized enterprises*, pp. 10–45. Routledge, 2022.

Dragomir, V. D., and M. Dumitru. "Practical solutions for circular business models in the fashion industry." *Cleaner Logistics and Supply Chain* (2022). https://doi.org/10.1016/j.clscn.2022.100040

Eberhardt, Leonora Charlotte Malabi, Morten Birkved, and Harpa Birgisdottir. "Building design and construction strategies for a circular economy." *Architectural Engineering and Design Management* 18, no. 2 (2022): 93–113. https://doi.org/10.1080/17452007.2020.1781588

Forastero, G. "Resources, conservation & recycling advances circular economy in Andalusia: A review of public and non-governmental initiatives." *Resources, Conservation & Recycling Advances* (2023). https://doi.org/10.1016/j.rcradv.2023.200133

Gupta, A. K. "Rare earth element extraction from secondary source materials: Understanding economics and environmental impacts of development." (2023). https://www.proquest.com/openview/3482db275b1e866ec2639a33accb8d8a/1?pq-origsite=gscholar&cbl=18750&diss=y

Hagelüken, C., and D. Goldmann. "Recycling and circular economy—towards a closed loop for metals in emerging clean technologies." *Mineral Economics* (2022). https://doi.org/10.1007/s13563-022-00319-1

Hegab, Hussien, Navneet Khanna, Neamat Monib, and Amr Salem. "Design for sustainable additive manufacturing: A review." *Sustainable Materials and Technologies* 35 (2023): e00576. https://doi.org/10.1016/j.susmat.2023.e00576

Jaakkola, E. "Designing conceptual articles: four approaches." *AMS Review* (2020). https://doi.org/10.1007/s13162-020-00161-0

Jeanrenaud, A., and P. Romanazzi. "Software product evaluation metrics: a methodological approach." *WIT Transactions on Information and Communication Technologies* 9 (2024). https://www.witpress.com/Secure/elibrary/papers/SQM94/SQM94005FU2.pdf

Khorasani, Mahyar, AmirHossein Ghasemi, Bernard Rolfe, and Ian Gibson. "Additive manufacturing a powerful tool for the aerospace industry." *Rapid Prototyping Journal* 28, no. 1 (2022): 87–100. https://doi.org/10.1108/RPJ-01-2021-0009

Kumar, A., P. Kumar, R. K. Mittal, and H. Singh. "Printing file formats for additive manufacturing technologies." *Advances in Additive Manufacturing* (2023). https://doi.org/10.1016/B978-0-323-91834-3.00006-5

Lamba, Pooja, Dilraj Preet Kaur, Seema Raj, and Jyoti Sorout. "Recycling/reuse of plastic waste as construction material for sustainable development: A review." *Environmental Science and Pollution Research* 29, no. 57 (2022): 86156–86179. https://doi.org/10.1007/s11356-021-16980-y

Mallick, Pravin Kumar, Kim Bang Salling, Daniela CA Pigosso, and Tim C. McAloone. "Closing the loop: Establishing reverse logistics for a circular economy, a systematic review."

Journal of Environmental Management 328 (2023): 117017. https://doi.org/10.1016/j.jenvman.2022.117017

Marsh, A. T. M., Velenturf, A. P. M., and Bernal, S. A. "Circular economy strategies for concrete: Implementation and integration." *Journal of Cleaner Production* (2022). https://doi.org/10.1016/j.jclepro.2022.132486

Moqri, Mahdi, Chiara Herzog, Jesse R. Poganik, Jamie Justice, Daniel W. Belsky, Albert Higgins-Chen, Alexey Moskalev et al. "Biomarkers of aging for the identification and evaluation of longevity interventions." *Cell* 186, no. 18 (2023): 3758–3775. https://doi.org/10.1016%2Fj.cell.2023.08.003

Neves, S. A., and A. C. Marques. "Drivers and barriers in the transition from a linear economy to a circular economy." *Journal of Cleaner Production* (2022). https://doi.org/10.1016/j.jclepro.2022.130865

Ostapska, Katarzyna, Petra Rüther, Arian Loli, and Klodian Gradeci. "Design for disassembly: A systematic scoping review and analysis of built structures designed for disassembly." *Sustainable Production and Consumption* 48 (2024): 377–395. https://doi.org/10.1016/j.spc.2024.05.014

Pacana, A., Siwiec, D., Bednárová, L., and Petrovský, J. "Improving the process of product design in a phase of life cycle assessment (LCA)." *Processes* (2023). https://www.mdpi.com/2227-9717/11/9/2579/pdf

Palladinetti, E. "When imitation isn't the sincerest form of flattery: Combating fast fashion design piracy through enhanced copyright protection." *Drexel Law Review* (2023). https://heinonline.org/HOL/LandingPage?handle=hein.journals/drexel15&div=14&id=&page=

Pichlak, Magdalena, and Adam R. Szromek. "Linking eco-innovation and circular economy—a conceptual approach." *Journal of Open Innovation: Technology, Market, and Complexity* 8, no. 3 (2022): 121. https://doi.org/10.3390/joitmc8030121

Praveena, B. A., N. Lokesh, Abdulrajak Buradi, N. Santhosh, B. L. Praveena, and R. Vignesh. "A comprehensive review of emerging additive manufacturing (3D printing technology): Methods, materials, applications, challenges, trends and future potential." *Materials Today: Proceedings* 52 (2022): 1309–1313. https://doi.org/10.1016/j.matpr.2021.11.059

Rai, Pawankumar, Srishti Mehrotra, Smriti Priya, Edgard Gnansounou, and Sandeep K. Sharma. "Recent advances in the sustainable design and applications of biodegradable polymers." *Bioresource Technology* 325 (2021): 124739. https://doi.org/10.1016/j.biortech.2021.124739

Raj, Tirath, K. Chandrasekhar, A. Naresh Kumar, and Sang-Hyoun Kim. "Lignocellulosic biomass as renewable feedstock for biodegradable and recyclable plastics production: A sustainable approach." *Renewable and Sustainable Energy Reviews* 158 (2022): 112130. https://doi.org/10.1016/j.rser.2022.112130

Reslan, M., N. Last, N. Mathur, K. C. Morris, and V. Ferrero. "Circular economy: A product life cycle perspective on engineering and manufacturing practices." *Procedia CIRP*, 105 *(*2022*)*: 851–858. https://doi.org/10.1016/j.procir.2022.02.141

Schepler, Xavier, Nabil Absi, and Antoine Jeanjean. "Refurbishment and remanufacturing planning model for pre-owned consumer electronics." *International Journal of Production Research* 62, no. 7 (2024): 2499–2521. https://doi.org/10.1080/00207543.2023.2218942

Stefanakis, Alexandros I., Cristina S. C. Calheiros, and Ioannis Nikolaou. "Nature-based solutions as a tool in the new circular economic model for climate change adaptation." *Circular Economy and Sustainability* 1 (2021): 303–318. https://doi.org/10.1007/s43615-021-00022-3

Tripathi, Shubham, and Manish Gupta. "Transforming towards a smarter supply chain." *International Journal of Logistics Systems and Management* 36, no. 3 (2020): 319–342. https://doi.org/10.1504/IJLSM.2020.10019293

Ullah, Mehran, Iqra Asghar, Muhammad Zahid, Muhammad Omair, Ali AlArjani, and Biswajit Sarkar. "Ramification of remanufacturing in a sustainable three-echelon closed-loop supply chain management for returnable products." *Journal of Cleaner Production* 290 (2021): 125609. https://doi.org/10.1016/j.jclepro.2020.125609
Velenturf, A. P. M., and P. Purnell. "Principles for a sustainable circular economy." *Sustainable Production and Consumption* (2021). https://doi.org/10.1016/j.spc.2021.02.018
Whalen, C. J., and K. A. Whalen. "Circular economy business models: A critical examination." *Journal of Economic Issues* (2020). https://doi.org/10.1080/00213624.2020.1778404
Xie, Zhenyang, Jian Wang, Guangyao Zhao, Qinyi Zhang, Hua Fan, Aohua Zeng, and Wei Ding. "A review on the durability of key components of PEM fuel cells." *Catalysis Science & Technology* (2024). https://doi.org/10.1039/D4CY00351A

9 Circular Business Models and Value Creation

Tran Minh Tung

9.1 INTRODUCTION

Recognize the societal integration of business. Your service or product's promise attracts customers. Moreover, those who produced it, supported it, and transported it, your files, and the natural environment all are part of the same success cycle. These different stakeholders represent inner, outer, and shared rings of context to services and goods. These service or product promises cannot be delivered without involving the service or product users, customers, support assets, prior relationships, and specific natural contexts. Economic prosperity and wealth are configured throughout this multiple ring of relationships that encompass what services and items provide today and explore implications for value enhancement, innovation, and business prosperity tomorrow. These intertwined social and environmental responsibilities underline a holistic approach to service and product businesses that are called circular business models (Awan & Sroufe, 2022; Salvador et al., 2021).

Given the accelerating consumption and wasteful disposal of resources, apply circular business models to increase business value creation and decrease negative environmental and social impact. Circular models produce significantly more environmentally beneficial and socially responsible economic prosperity that does not depend on planetary resource sacrifice and equity depletion (Tapaninaho & Heikkinen, 2022). Circular economy pioneers extract maximum value from materials and regenerate depleted environments. Therefore, whether you are a front-runner or just starting on a circular journey, accelerate your transition toward circular business models that build, earn, and consistently provide long-term, refreshing, and renewable business prosperity that is built, earned, received, and reciprocated from the core social and environmental values of the service and product solution. Focus on how to discover and improve your business priorities, relationships, and assets that support circular drivers that buffer and extend significant circularity capabilities (Hart, 2023).

9.1.1 Background and Rationale

To have an impact, new CE business models need to support an innovative culture. They should stimulate organizations and circle members to stay active, and they should push for continuous interaction. This enables development, learning, and knowledge exchange—all of which are vital for achieving systemic change. New and

DOI: 10.1201/9781003538837-9

innovative business models accentuate the importance of collaboration and co-development. They aim to connect different companies and stakeholders in a cooperative circular business network that can accelerate the transition to CE (Dokter et al., 2023).

In order to realize potential savings by connecting circular economy (CE) strategies with product development, companies must redesign their operations and business models. Moving from selling products to making them available as a service is a significant shift. It requires a change in the organization of production and consumption processes and also leads to a change in value creation. Furthermore, the resulting transformation in industry also stimulates the exploration of new trajectories for local and regional economic development. All of these trajectories must be deeply rooted in society, the environment, and the well-being of all citizens. In other words, we need new business models for circular economy (Gebauer et al., 2021).

9.1.2 Research Objectives

At this point, companies can be inspired by the various existing closed-loop economy models in order to ensure the sustainable use of the resources they use and to minimize damage to the environment with the products, services, and processes they provide. Circular business models are models used by companies to integrate environmental, social, and ethical interests of the economy for the purposes of opening new revenue streams and protecting existing revenues, as well as reducing the damage to the environment and providing new benefits to customers. Circular business models primarily aim to create new sources of revenue and value creation for companies (Kanda et al., 2021). In addition to the goal of value creation, companies need to adopt and use these models to create a broader perspective for the creation and capture of value on multiple levels, effectively on an ecological, social, financial, and business level (Fehrer & Wieland, 2021).

The rapid changes in economic, social, environmental, political, and technological dimensions have recently begun to alter the nature of business. These changes encourage and force companies to integrate environmental, social, and ethical interests into traditional business models. In the same way, customer needs, demands, and behaviors also began to change in parallel with these changes. Consumers are looking for more affordable, high-quality, practical, ethical, and environment-friendly products. Providing products or services that meet these needs has become one of the main objectives in the process of creating new revenues and sustaining existing revenues for organizations (Sarkar et al., 2022).

9.2 CONCEPTUAL FRAMEWORK

Circular business models have gained attention from academia and business practice due to the urgency of sustainable development and the realization that circularity could support the realization of that urgency. It is no wonder that, together with widespread public advocacy for circular economy, industry actors come with practical initiatives in energy, building, food, consumer goods, and more down the value input and end-of-life recycling chain (Hina et al., 2022). Examples are Ford moving toward a fully circular business model using remanufacturing principles, the Ellen MacArthur Foundation through multiple initiatives to build and scale circular-friendly business models. Besides these fragmented efforts, often called best practice explorations or pilots, few researchers design interorganizational structures that stimulate multiple businesses in the same value chain to design circular business models together. Yet, in today's circumstances, societal pressures beyond political borders have not yielded sufficient exploration through business model design, in line with increasing criticism of not moving forward to new business structures on EU levels and theoretical criticism for overemphasizing generic circular business principles (Aguilar Esteva et al., 2021; Demartini et al., 2023).

This chapter is motivated by the need to better understand how circular business models (CBMs) can be developed to realize value creation for businesses and the customers they serve. In particular, our question is: How do CBMs account for design decisions that alter value creation processes and influence consumption models? Circular business models (CBM) come in different shades. Each of these paths changes the way companies approach the resource value cycle of reusing, reducing the consumption of, or recycling resources. The primary goal of circular business models is to drive corporate value creation (Ferasso et al., 2020). The resource cycles these business models use can be the water flows in food production processes or steel in a steel co-production system, the components and materials in a product delivery chain. Typically, the most valorized resources in the chain are those that circulate in smaller cycles, with a higher fraction of the resource recirculated (Kanda et al., 2021).

9.2.1 Circular Economy Principles

The first principle is the regenerative design principle, which ensures that products last as long as possible through electrification, enable the reuse of products, and move the use of biologically and technically compatible materials and components deeper into the production and supply chain. Second, it is the principle that focuses on the use of renewable resources during the production phase. In this way, the company will make the most of the raw materials offered by our planet, and avoid their waste by eliminating the incineration and disposal of materials in landfills. The reuse or recycling of materials will then take place, helping the company to promote and expand the capabilities of existing recycling infrastructure (Aguiar et al., 2022).

The circular economy aims to redefine the market where we eliminate waste and design products and services for better sustainability. The focus of the circular economy is to maintain the value that would be lost across the life cycle of our products and components. These principles require the implementation of long-term thinking in the entire structure of the business model while keeping an eye on the design of the product we are delivering as well as the stages of production that follow. The concepts of circular economy and business models need to be addressed together, in both the company and the state. Therefore, these principles support the development of clean and smart production processes, based on clean technologies. A company implementing the circular economy principles as part of its business model will develop a culture to maintain and promote the value of products and resources during their life cycle (Velenturf & Purnell, 2021; Corvellec et al., 2022).

9.2.2 Business Model Innovation

So, the question that arises is how companies can contribute to sustainable development on a more interrelating scale by increasing the capital of natural resources and doing so within business principles based on the organization's objectives of creating value. By using the principles of the circular economy, it is possible to change the traditional Business of companies and make them more circular (Salvador et al., 2021).

The Business Models traditionally provide the connection services between clients and companies. They describe the range of the product and services' value generated by the company, as well as the pricing system. It is composed of tools such as customers, value offers, expenses, competency, incomes, municipal system, margin, and competitive advantages. Since transaction cost economics focuses on understanding the different determination elements in firms' boundaries and on the market substitute, the property-based and the organization-based views both are concerned with how the business model is defined and implemented (Sjödin et al., 2020).

Sustainable development is anchored in several aspects, ranging from social and environmental aspects to economic issues. This concept has gained significant visibility worldwide with the Brundtland Report (Brundtland 1987). Fifteen years later, in 2002, The World Summit on Sustainable Development held in Johannesburg outlined a general program, expanding the concept to an international debate. In Brazil, the subject is addressed in the study of Corporate Sustainability of the Getulio Vargas Foundation (FGV—GVCes) which has been monitoring companies that consider the sustainability axis in their business management for eight years (Lozano, 2024).

9.3 LITERATURE REVIEW

The modern concept of "circular economy" emerged in the second half of the 20th century—earlier than the name: the first attempts to describe it were made during the oil crises of the 1970s. Over the past 50 years, the concept has undergone significant changes and transformations, which have led to the formation of the most complete form of circular economy, described by the Ellen MacArthur Foundation as a framework for an economy that is restorative and regenerative by its core design. This regenerative form of the circular economy requires a transition in all three segments of the economy: from a linear to a circular supply chain organization (where we look at the physical aspects of a supply chain), from a business model inherited from the industrial age to a less productive model that provides service-based solutions (where we focus on supplying customers with what they need), and from the point of view of system outcomes (having objectives that are consistent with the systems where they are under control) (Tuladhar et al., 2022).

9.3.1 Defining Circular Business Model in the Scope of Research

9.3.1.1 Circular Economy Theory

The concept of a circular economy has been attracting interest and attention from academic researchers studying the general literature. Over the years, the concept has been identified as a conceptual background in different scientific fields such as economics, corporate social responsibility, waste management models, sustainability policies, organizational management, and business models. The focus of the attention of various authors is the transition from a linear to a circular economy, the disruptions related to the transition, and the effort needed for the system to transform from a traditional economic system to a circular model. These scientific works can teach and prepare all actors, help avoid obstacles, and plan for a global transition based on economic, social, and environmental changes, while reducing impact and uncertainty. The circular economy model recalls the old model of traditional society, based on durables and reusable, and the much older model of nature itself, where everything can be reused and nothing is lost (Alcalde-Calonge et al., 2022; Centobelli et al., 2020).

The circular economy model has been widely recognized within academia and industry as an approach that combines economic development with environmental and social perspectives. Despite the circular economy having emerged from the concept of sustainable development, the model itself has its own identity, playing an important role in the economic development of the coming years. Evidence of continuity will be found through circular initiatives. The wastage of valuable resources and a fragile global financial system put the economy under pressure. Natural and financial capitals will be the focus of the transition. By valuing and conserving natural resources, capital becomes more robust and risk decreases, reducing the system's reliance on natural resources. Companies with a business model designed to create value in a circular economy realize the opportunities that healthy people and a planet offer, and they can review and innovate their designs, materials, and production flows (Massaro et al., 2021; Nikolaou et al., 2021).

9.3.2 Sustainable Business Models

Giving up existing consumption and production systems requires major economic and social transformation. In order to minimize the burden on economic and social transformation, organizations are in need of sustainable business models and the right trajectory that will implement and sustain those models. The sustainable business model is as crucial as the economic, social, and environmental approaches, bringing the processes necessary for the realization of the philosophy and objectives that underlie sustainability and combining them with strategy. However, in order to create a positive impact, sustainable business models require a new look at opportunities, processes, and methods focusing on the environment, society, and economy. In the context of this study, the sustainable management model consists of economy, environment, and social dimensions (Khan et al., 2021).

Years after natural resource exhaustion and environmental deterioration were reported as significant problems, organizations have realized that there is no trade-off between economic and ecological/environmental issues. In this vein, current organizational research is seeking new paths that will lead to organizations operating more sustainably. Emerging from these concerns, the concepts of circular economy and business models are receiving particular attention today. Circular business models that are responsible for the use of scarce resources and creating value in accordance with economic, environmental, and social dimensions can be widely adapted to different sectors (Table 9.1) (Mostaghel & Chirumalla, 2021).

TABLE 9.1
A Framework for Understanding Circular Business Models

No.	Section	Description	Authors, Sources
1	Background	• Growing pressure for sustainable development • Circular economy offers solutions • Businesses are starting to implement circular models • Need for more research on designing these models collaboratively	Aguiar et al. (2021), Demartini et al. (2023), Ferasso et al. (2020)
2	Circular economy principles	• Regenerative design: durable products, reuse, biocompatible materials • Renewable resources • Reusing and recycling materials	Aguiar et al. (2022), Velenturf and Purnell (2021), Corvellec et al. (2022)
3	Business model innovation	• Traditional versus circular business models • Circular models focus on value creation throughout a product's life cycle	Salvador et al. (2021), Sjödin et al. (2020)
4	Literature review	• Circular economy concept has been evolving for decades • Importance of transitioning to a circular economy across all levels (supply chain, business model, systems)	Tuladhar et al. (2022)
5	Defining circular business model	• Circular economy theory is a foundation for circular business models • Businesses can benefit economically, socially, and environmentally from circular models	Alcalde-Calonge et al. (2022), Centobelli et al. (2020), Massaro et al. (2021), Nikolaou et al. (2021)
6	Sustainable business models	• Sustainable business models are necessary for a smooth transition to a circular economy • They consider economic, social, and environmental aspects	Khan et al. (2021), Mostaghel and Chirumalla (2021)

Source: Synthesized by author.

9.4 METHODOLOGY

In the next phase, a set of entrepreneurship applications have been used mainly by two companies and by the researchers, who sought to mobilize, facilitate, and stimulate research activities and the development of the companies taking part in CIR-C4Life (Purola et al., 2020), (Verleye et al., 2024). At this juncture, two roundtable initiatives counted on the participation of several companies to support the construction of the circular business models. This roundtable was important to validate and further refine the findings obtained in previous phases. The model was submitted to open consultations of the European Circular Business Model Project. During this consultation, stakeholder workshops have been completed in five different European countries. In the last phase of the action-research, the model was tested in 12 new organizations, asking for a web-based questionnaire that represents the ideal of the business model according to respondents' point of view. The current version of the visual model is the result of action-research and is a proposal for business model typology and components.

The investigations of the topic have soared in the last decade, and several studies, frameworks, and practical guidelines are available to help understand the current practices and develop new business models and processes. The presented model was influenced by the research that was realized in the CIR-C4Life, CIRCit, and CIRC-SEDE projects (Fehrer & Wieland, 2021). The programs are financed under the Horizon 2020 research and innovation program. These programs aim to develop a platform to measure, monitor, and control the bioeconomy and the circular economy model. Other studies and particularly the perspectives of industrialists and entrepreneurs that add richness and specificity to the model have further expanded these analyses. The business model component was developed through action-research and workshops, using as a starting point Bisgambiglia & Bellot's Typo Base générale de business model, that is often used by the French OPECST to provide a basic categorization of different types of business models. Two different questionnaires were developed and tested through a pretest in two companies. The obtained data was analyzed using SPSS to identify which components are considered more important by the participating organizations.

9.4.1 Research Design

This study is empirical in that it collects text on government policies, corporate circular economy strategies, historical business models of companies, and other research studies on business and economic theories. Since the study does not rely on primary data collection, it is not a field study. The fact that the study is industry-based can be seen as archival-based research. This is compensated for by the use of mixed methods (Bouteraa, 2024). Data of both a quantitative and a qualitative type are sought, along with data obtained from government reports, academic studies, and books. The result is that the study is high in construct validity. The deductive method will use logic and guidance from well-known economic theories on what those "particular pieces of information" that matter could be.

The chosen research methodology for this study can best be described as qualitative, exploratory, and descriptive, applying both an inductive and a deductive approach. Qualitative research is particularly useful for conceptual investigations, providing the researcher with an insight into those concepts, beliefs, and behaviors as perceived by the researched subjects (Pope & Mays, 2020). However, the specific research question of this study, "How can companies identify, develop, and exploit business models that circulate resource efficiency for the purpose of circular economy?" is not directly answerable by sphere-obtained quantitative data tracing effects or linkages in the economic world. Analytical or logical induction, however, means producing generalizations from particular instances, thus combining data and existing theory. The research process then tries to establish robustness for emerging theory using both establishing construct validity and conducting more tests.

9.4.2 Data Collection and Analysis

The research method used in this chapter includes evaluating several business cases and business model descriptions. The aim of the assessment is to detect primary business models that support circularity concepts. The constructed list of business models is not to be considered as completed but rather as needing clarification or adjustments. Adjustment would include retiring some business models, collapsing some business models, changing the definition of some business models, and adding new business models. Circular business models that may need further research to be state-of-the-art models are considered innovative. Gathering business models and offering legal entities specific support could contribute to the commonly needed business modeling architecture with rules, norms, and standards (Kanda et al., 2021).

Several data sources are used to identify primary circular business models. For sustainable decision-making in terms of circularity, assessing the quality of various sources of data is essential. This leads to the consideration of indicators used to measure the business models. It is also important to assess the maturity of the structures and the functionalities that support the implementation of certain business models. This assessment brings complementary qualitative information during the operational analysis of models and structures. The final section of this chapter presents results and findings of the research.

9.5 CASE STUDIES

In addition, several enabling technologies for waste and pollution management have emerged in recent years. Traditional industries are also seeking to explore shared asset use and return systems. In addition, the industrial symbiosis lifestyle often has similar attributes to product-service systems. In the fourth and fifth sections, a variety of policy and design tools are used to develop strategies and intrinsic and systemic business models, including industrial and service integration. These proposals can boost sustainable value creation, service improvement, and service enhancement possibilities in both products and circular and circular cities. Summaries of cases are provided in Table 9.2. These case studies highlight various industry sectors and showcase how the circular economy is applied in different ways.

These case studies are based on different industry sectors and illustrate the application of circular economy to policymaking, product design, service business models, and regional integration. For example, governments in various countries and regions support manufacturers, system integrators, and service providers through issuing guidelines, providing financial assistance, undertaking demonstration projects, and taking the lead in using circular products and services. The emergence of a systemic transition plan will lead to considerable demand for labor skills. Such incentives can significantly improve resource efficiency by creating skilled jobs for effective resource recovery, energy efficiency, and material and product redesign.

TABLE 9.2
Case Studies of the Circular Economy in Action

Sector	Location	Example	Circular Economy Application	Source
Construction and waste	Vietnam (Hanoi)	Greenpeace's "Eco-bricks" Project: Turning nonrecyclable plastic waste into building blocks for schools and community centers	Product Design & Waste Management	Greenpeace Eco-bricks Project Vietnam on greenpeace.org
Fashion	EU (The Netherlands)	Patagonia's Worn Wear Program: Repairing, refurbishing, and reselling used Patagonia clothing, extending product lifespan	Product Design & Service Business Model	Patagonia Worn Wear Program on patagonia.com
Food	USA (California)	Imperfect Foods: Rescuing cosmetically imperfect or surplus produce from farms, reducing food waste	Waste Management & Service Business Model	Imperfect Foods website on imperfectfoods.com
Electronics	Asia (Japan)	Panasonic's "Refurbished Products" Initiative: Offering refurbished electronics at discounted prices, extending product life and reducing e-waste	Product Design & Service Business Model	Panasonic Refurbished Products website on Panasonic
Agriculture	Vietnam (Mekong Delta)	The Organic Rice Project: Farmers using rice straw for mushroom cultivation, creating additional income, and reducing agricultural waste	Resource Management & Regional Integration	The Organic Rice Project
Packaging	EU (Germany)	REWE Group's Reusable Packaging System: Offering reusable containers for fruits and vegetables, reducing single-use plastic packaging	Policymaking & Product Design	REWE Group Reusable Packaging System Press Release on presseportal.de
Manufacturing	USA (Michigan)	Ford's "Redesign for End-of-Life Vehicles" Program: Designing cars for easier disassembly and parts reuse in future models	Product Design & Policymaking	Ford Redesign for End-of-Life Vehicles Program Press Release on media.ford.com

Source: Synthesized by author.

9.5.1 Company A: Sustainable Supply Chain Management

When recycling polymers, she/he pays typically about 1/10 for new fibers for a replacement of the polymers made of PA6 fibers and therefore can replace the polymer four times before costs are added up. The CO_2 emissions of their simple replacing process are three times the CO_2 of the rematerialization process. When replacing this performance with the production of new polymers, he needs to pay an additional 30 EUR/kg and throughput 4 polymers of a weight of 1,000 kg through the industrial processes for making the peak of the polymers. Then, the company has to replace the polymers three times as well to fulfill the same requirements of his final product. Thus, the cost for polymers and emissions are three times lower than in the simple replacement process (Subramani et al., 2024; Law & Narayan, 2022).

This demonstrates the potential of corporate decision-makers for optimizing their business models. The company A case shows the use of a closed-loop recycling model. Closed-loop recycling is the highest value of all recycling processes and therefore company A walks the talk when saying that it uses the highest available recycling processes. Because company A operates in a very competitive business area, their profit is highly symmetric with higher quantities of recycled material. Their service costs are linear and also costs of material compensation are proportional. Together with the fact that simple replacement creates three times the costs and creates three times the emissions, their business is ruled by profits and the environment.

9.5.2 Company B: Product Life Extension Strategies

The case of Business B shows an interesting hybrid combination of circular business model characteristics that extend the life of a product, and subsequently obtain additional value from the product or its materials (primarily driven by product life extension opportunities for a product). On the one hand, the highest priority company value that is ascribed to their wash barrels is the non-product-related value of being able to ease the company's relationship with their customers in dealing with used textiles, while also raising awareness around corporate social responsibility. On the other hand, the wash barrel is definitely a standout example of an innovative product that drives a small revenue while the washing service domain related to the wash barrel consists of tenants being able to use the barrels as washing machines; Placelet is able to form another method of revenue generation, building upon existing relationships (Staples & Krumel Jr, 2023; Back et al., 2021).

In the context of the product life extension strategies pursued by the two case companies, Company B's strategies align more closely with the high involvement and high engagement propositions. High-involvement strategies imply a deep and embedded commitment on the part of the company to realize the value from the extension strategy, and potentially involve one or more of the external stakeholders. High-engagement strategies imply high levels of co-creation of value with external stakeholders. Company B's approach is more collaborative with a broader spectrum of stakeholders since they engage with existing waste contractor firms and social enterprises to deliver the wash barrels and to create craft-based and machine-based handicrafts from garments. The go-to-market circular strategies deployed to drive the wash barrel sales are currently more conventional.

9.6 FINDINGS AND DISCUSSION

First, the chapter has listed 11 archetypes of circular business models. These archetypes contain layers of innovation, which allow businesses to resonate with the aims of the circular economy by adhering to different configurations of the business model, that is, the way in which the organization interacts with the external and internal business environment. The business environment is analyzed as a complex and adaptive system, looking through the stakeholder perspective and considering different levels of analysis. Each archetype can suggest concrete avenues for creating economic, ecological, and social values, adding to a firm's competitive advantages (Salvador et al., 2023).

This chapter has reviewed 56 contributions in the field of circular business models to identify the circular business model archetypes and the inducement mechanisms to business model innovation that arise from adopting circular business models. Additionally, this research has also considered areas where managers can act to protect or create commercial value through the circular business models they design and execute. This provides a framework and foundation to advance research into circular business models and value creation under a systems thinking approach and different levels of analysis.

9.6.1 Findings

9.6.1.1 Key Success Factors in Circular Business Models

All businesses, and even cities and regions, should consider rethinking the predominantly linear structures of their business models, looking for benefits and opportunities. It is, therefore, a challenging problem for all entities concerned with issues of sustainable development. Governments are forced to take a significant role in this transition through regulations, new economic instruments, or specific public policies, emphasizing the incentive structures. Any of these measures, together or separately, can help to create an economic system that values, actualizes, or even finances its potential connections, and thus ensures environmental, socioeconomic, and financial profitability. Circular economic systems presuppose that interventions and decisions be taken following the guided long-term vision by any of these political sectors and that the general public flows in the same direction to create a competitive advantage (Iacovidou et al., 2021).

Circular business models (CBMs) propose quite different profit formulas than type 1 and type 2 business models. These are still based on value creation, but on a much broader concept. Their economic and environmental viability depends widely on the existence of attributes and organizational capabilities, so their main focus is success in these models. The description of these features and capabilities is essential both for the theoretical consolidation of the research conducted on CBMs and, above all, to make them practically relevant. In fact, the environmental and social problems associated with the scarcity and cost of raw materials, the excess of landfills, and the high pollution rates depend on the urgency to change from linear business models to CBMs (Fehrer & Wieland, 2021; Averina et al., 2022).

9.6.2 Challenges and Limitations

Despite the potential of circular business models, they also pose nontrivial challenges. All circular business models require complex operational mechanisms and adequately skilled labor pools to maintain the flow of products and components, and the quality/certification issues are also relevant (e.g., to increase the acceptance of reused or recycled in new products). In addition, certain challenges have been identified for each of the three main types: while PSS-based and RBC-based business models require critical internalization of the costs of resource recovery and reuse, the RLC-based business models may require performers to make large investments in productions for a perceived higher risk compared to traditional models (Hina et al., 2022). Furthermore, there is still a limited availability of guidelines and tools addressing how circularity requirements can be incorporated into firms' operations, supply chain, and networks, the choice of the remanufacturing strategy, and the limits of standardization. However, some initial steps toward the development of appropriate capabilities, based on enhancing openness to knowledge from external networks, de-siloing of relevant internal knowledge that may be dispersed around different departments/units, and the changes required for workers, as well as potential barriers to such changes, have been recorded. Finally, it is crucial to define the best timing for reuse activities, while considering the age, history, and maintenance costs of obsolete goods awaiting their next life. The quantitative and qualitative importance of these limitations very much depends on the specific characteristics of the products/services to be reused and recycled, the nature of the customer and their needs, and so on, but together, they suggest a significant attitude toward trying feasible restarts (Salvador et al., 2020).

9.7 IMPLICATIONS FOR PRACTICE

One key argument given in business analysis to not exploit circular principles is that of the expected costs, especially during initial steps. What impacts does the extending of these responsibilities to the other two loops provided by the framework have? Of those interviewed during our research, the majority saw the innovation costs for CBM in many cases as feasible. Even actors in procurement of engines thus perceive the benefits of efficiency in use. Understandably, the transition to CE is not easy nor useful in all parts of the enterprise. Such flexibility demonstrates the theoretical potential for diverse strategies for the practical organizational capabilities that could make value creation more sustainable. However, it could produce more sustainable value in specific settings.

The practice implications in the following are derived from the findings in Chapter 5 and our interpretation of what circular CBMs involve. Suggesting the implementation of CBM opens up opportunities, not least because when enterprises add another layer of functioning to one that is already effective, efficiency increases. This means that there is potential for a win-win scenario. The CBMs create opportunities for increasing resource efficiency or for generating direct savings and the process capabilities of the enterprise by providing a basis for higher revenues (Dembek & York, 2022). A first important consideration is whether implementing CBMs in business contexts could actually occur since the motivations and the organizational practices needed diverge from conventional management principles. To clarify, we offer possible management activities.

9.7.1 Recommendations for Companies

Small- and medium-sized enterprises (SMEs) dominate EU economies. Whereas a certain percentage of both small- and medium-sized enterprises engage in forward-thinking practices aligned with the principles of the circular economy (CE), the circular business model (CBM) literature is imbalanced and dominated by studies that concern only large firms (García-Quevedo et al., 2020). SMEs have unique characteristics and might require different CBMs. What are some CBMs suitable for SMEs? Moreover, what are potential challenges and solutions that SMEs might encounter when exploring and implementing them? The aim of this chapter is to address these research questions through an exploratory review of the circular business model literature followed by a thematic and content analysis. This examination results in a typology that categorizes the different CBMs to emerge based on their underlying circularity building blocks and illustrates their unique value creation logic. We then attempt to match these CBM types to the unique characteristics of SMEs. The chapter results in a multifaceted perspective on CBMs and align SMEs to those CBM types that fit their unique characteristics best.

Implementing multiple business model circularity pathways concurrently, for example, through an integrated product-service-system (IPSS) approach, can increase both resource productivity and business model resilience. Companies of all sizes need to develop new business models and complementary strategies designed for a circular economy (Zhang et al., 2024). However, while the opportunities accompanying circular business models are large, there are only a few CBMs and there is often little empirical and even fewer quantitative research findings or insights for developing businesses. Thus, the question arises: What could companies that explore CE-inspired business models do and what might others learn from these experiences? Recommendations are to (1) attempt to establish local ecosystems for a CBM; (2) focus on reverse salients when trying to foster eco-behavior change; (3) make boundary resources available for circular citizens; (4) not focus on the direct pollution effect of private household spending but on its consistency instead, and (5) explore implementers' myths of circularity.

9.8 CONCLUSION AND FUTURE RESEARCH DIRECTIONS

Given the high and increasing interest in circular business models within the scientific and practice community and the broad, interdisciplinary requirements for supporting the actualization of circular business models, the identified research avenues are only a starting point (Fehrer & Wieland, 2021). First, the renewable material concentration potentially induces new manufacturing techniques and a strong demand for new materials. The fast pace of technological adaptation puts a focus on more rapid and distributed technological innovation processes. Second, the new waste facilitating business models, process, and ethics, which resemble more horizontal or even democratic decision-making structures. The third aspect related to business model design focuses on a more hybrid approach with democratizing elements in their roles as elements in this business logic. Through these implications, our chapter suggests further research directions on circular business model properties, related process aspects, their legitimacy, and power relations.

The aim of this chapter is to provide an overview and integrate the current state of knowledge on circular business models and value creation, and to structure the insights gained. We have shown that business models are central to realizing the circular economy and that there is a diversity of strategies and models used within the circular economy. To also account for the fact that circular business models provide value creation at various levels and scales and are supported by a unique set of implementation factors ranging from corporate strategic positioning to non-corporate collaboration, we apply the CAMA model through a thematic cluster analysis to make these insights more accessible to the academic and practitioner community. The challenges of developing models for value components, framing circular business model contributions within the larger discussion of transforming economic systems, and dynamics of success or failure are identified as important future research avenues (Goni et al., 2021).

9.8.1 Summary of Findings

Another key finding is that firms that link more than one business model experience have greater leverage of their intangibles in improving both their financial and socio-environmental performances. Therefore, integrating a new business model in established ecosystems creates synergies that enhance both personalized service provision and corporate sustainability. This study reveals that placing emphasis on a new business model reliant on customers' ability to handle some service functions downloads the competitive advantages specific to engaging in one single business model. Management education could benefit from targeted discussion addressing these shared outcomes. Finally, the results show that the trade-offs in cost dilution of firms that engage in multiple business models and try to shift from economic to environmental performance can be reduced (Laguir et al., 2024). This scenario emphasizes the strength of the relationship between earnings management and environmental disclosure. The chapter ends with implications for research and practice.

This study expands the circle economy discussion by examining how firms' different business models influence the relationship between a firm's financial performance and corporate sustainability. We attempt to address the paucity of studies that consider all business models engaged in a firm, which highlights the need to study the simultaneous fit of a diversified set of business models to unlock economic and environmental performance improvements. The results suggest that mixed business model effects impact the relationship between economic performance and a firm's corporate social performance (Al-Shammari et al., 2022). Therefore, firms that subscribe to more than one business model need to commit to a broader range of sustainability initiatives if they wish to enhance their financial performance. Commercial rental companies need to develop the capabilities to manage operations that require an e-business model. This suggestion may seem logical if firms from this sector were to claim that the services they provide are environmentally less harmful than the ones produced by pure-play retailing companies. We believe that this response is not necessarily true, that is, while products do not change, services hold the potential to denote the differentiation linked to circularity.

9.8.2 Areas for Further Exploration

We do this by further refining our research model introducing functionalities of the different levels. The upper level in our research model is represented by the circular economy. In previous sections, we have already illustrated that business models define the way companies run their organization. Companies have a set of core capabilities that are interrelated and underlying capabilities enable their primary capabilities. Both types of capabilities are deployed and generate level outcomes. These outcomes, together with other factors such as the financial environment (e.g., investments, costs, income), the market (e.g., demand and supply), and the government (laws and regulations), provide the value creation that we want to discover. Our interest is in the different combinations of the three levels and how companies can leverage the functionality and interaction of these levels to generate value.

What is it that makes these 15 building blocks of circular business models come together in such a way that there are connections and flows of goods, data, and assets? And how does this implementation and the resulting circular model actually drive value creation? These are questions that our model cannot cover at present but we believe are crucial to unlock the ambition that is in the shift toward a circular economy. Thus, these questions form fertile ground for future research. In the following sections, we will address each of the three levels (the economy, the enterprise, and the offering level) with ideas representing potential research topics to explore where value is created as the building blocks combine into a full circular model. Before we explore opportunities for further research, it is beneficial to look back at the structure of our research and the functioning of our business model.

REFERENCES

Aguiar, M. F., Mesa, J. A., Jugend, D., Pinheiro, M. A. P., & Fiorini, P. D. C. (2022). Circular product design: Strategies, challenges and relationships with new product development. *Management of Environmental Quality: An International Journal*, 33(2), 300–329. https://doi.org/10.1108/MEQ-06-2021-0125

Aguilar Esteva, L. C., Kasliwal, A., Kinzler, M. S., Kim, H. C., & Keoleian, G. A. (2021). Circular economy framework for automobiles: Closing energy and material loops. *Journal of Industrial Ecology*, 25(4), 877–889. https://doi.org/10.1111/jiec.13088

Al-Shammari, M. A., Banerjee, S. N., & Rasheed, A. A. (2022). Corporate social responsibility and firm performance: A theory of dual responsibility. *Management Decision*. https://doi.org/10.1108/MD-12-2020-1584

Alcalde-Calonge, A., Sáez-Martínez, F. J., & Ruiz-Palomino, P. (2022). Evolution of research on circular economy and related trends and topics. A thirteen-year review. *Ecological Informatics*, 70, 101716. https://doi.org/10.1016/j.ecoinf.2022.101716

Averina, E., Frishammar, J., & Parida, V. (2022). Assessing sustainability opportunities for circular business models. *Business Strategy and the Environment*, 31(4), 1464–1487. https://doi.org/10.1002/bse.2964

Awan, U., & Sroufe, R. (2022). Sustainability in the circular economy: Insights and dynamics of designing circular business models. *Applied Sciences*. https://doi.org/10.3390/app12031521

Back, R. M., Lowry, L. L., & Higgins, L. M. (2021). Exploring a wine farm micro-cluster: A novel business model of diversified ownership. *Journal of Vacation Marketing*, 27(1), 103–116. https://doi.org/10.1177/1356766720954258

Bouteraa, M. (2024). Mixed-methods approach to investigating the diffusion of FinTech services: Enriching the applicability of TOE and UTAUT models. *Journal of Islamic Marketing*. https://doi.org/10.1108/JIMA-12-2022-0343

Brundtland, G. H. (Chair of the World Commission on Environment and Development). (1987, October). *Our Common Future: Report of the World Commission on Environment and Development* (UN Document A/42/427). Oxford University Press. 383 pp.

Centobelli, P., Cerchione, R., Chiaroni, D., Del Vecchio, P., & Urbinati, A. (2020). Designing business models in circular economy: A systematic literature review and research agenda. *Business Strategy and the Environment*, 29(4), 1734–1749. https://doi.org/10.1002/bse.2466

Corvellec, H., Stowell, A. F., & Johansson, N. (2022). Critiques of the circular economy. *Journal of Industrial Ecology*, 26(2), 421–432. https://doi.org/10.1111/jiec.13187

Demartini, M., Ferrari, M., Govindan, K., & Tonelli, F. (2023). The transition to electric vehicles and a net zero economy: A model based on circular economy, stakeholder theory, and system thinking approach. *Journal of Cleaner Production*, 410, 137031. https://doi.org/10.1016/j.jclepro.2023.137031

Dembek, K., & York, J. (2022). Applying a sustainable business model lens to mutual value creation with base of the pyramid suppliers. *Business & Society*. https://doi.org/10.1177/0007650320973450

Dokter, G., Boks, C., Rahe, U., Jansen, B. W., Hagejärd, S., & Thuvander, L. (2023). The role of prototyping and co-creation in circular economy-oriented innovation: A longitudinal case study in the kitchen industry. *Sustainable Production and Consumption*, 39, 230–243. https://doi.org/10.1016/j.spc.2023.05.012

Fehrer, J. A., & Wieland, H. (2021). A systemic logic for circular business models. *Journal of Business Research*. https://doi.org/10.1016/j.jbusres.2020.02.010

Ferasso, M., Beliaeva, T., Kraus, S., Clauss, T., & Ribeiro-Soriano, D. (2020). Circular economy business models: The state of research and avenues ahead. *Business Strategy and the Environment*, 29(8), 3006–3024. https://doi.org/10.1002/bse.2554

García-Quevedo, J., Jové-Llopis, E., & Martínez-Ros, E. (2020). Barriers to the circular economy in European small and medium-sized firms. *Business Strategy and the Environment*, 29(6), 2450–2464. https://doi.org/10.1002/bse.2513

Gebauer, H., Paiola, M., Saccani, N., & Rapaccini, M. (2021). Digital servitization: Crossing the perspectives of digitization and servitization. *Industrial Marketing Management*, 93, 382–388. https://doi.org/10.1016/j.indmarman.2020.05.011

Goni, F. A., Gholamzadeh Chofreh, A., Estaki Orakani, Z., Klemeš, J. J., Davoudi, M., & Mardani, A. (2021). Sustainable business model: A review and framework development. *Clean Technologies and Environmental Policy*, 23, 889–897. https://doi.org/10.1007/s10098-020-01886-z

Hart, A. (2023). Circular economy: Closing the catalyst loop with metal reclamation from spent catalysts, industrial waste, waste shells and animal bones. *Biomass Conversion and Biorefinery*. https://doi.org/10.1007/s13399-021-01942-8

Hina, M., Chauhan, C., Kaur, P., Kraus, S., & Dhir, A. (2022). Drivers and barriers of circular economy business models: Where we are now, and where we are heading. *Journal of Cleaner Production*, 333, 130049. https://doi.org/10.1016/j.jclepro.2021.130049

Iacovidou, E., Hahladakis, J. N., & Purnell, P. (2021). A systems thinking approach to understanding the challenges of achieving the circular economy. *Environmental Science and Pollution Research*, 28, 24785–24806. https://doi.org/10.1007/s11356-020-11725-9

Kanda, W., Geissdoerfer, M., & Hjelm, O. (2021). From circular business models to circular business ecosystems. *Business Strategy and the Environment*, 30(6), 2814–2829. https://doi.org/10.1002/bse.2895

Khan, I. S., Ahmad, M. O., & Majava, J. (2021). Industry 4.0 and sustainable development: A systematic mapping of triple bottom line, Circular Economy and Sustainable Business Models perspectives. *Journal of Cleaner Production*. https://doi.org/10.1016/j.jclepro.2021.126655

Laguir, I., Stekelorum, R., Beysül, A., Mouadili, O., & Segbotangni, E. A. (2024). Eco-control systems and firm performance: Understanding the mediating role of circular economy practices and the moderating role of environmental uncertainty. *Journal of Cleaner Production*, 450, 141894. https://doi.org/10.1016/j.jclepro.2024.141894

Law, K. L., & Narayan, R. (2022). Reducing environmental plastic pollution by designing polymer materials for managed end-of-life. *Nature Reviews Materials*. https://doi.org/10.1038/s41578-021-00382-0

Lozano, R. (2024). Sustainable development and sustainability. In: *Organisational Change Management for Sustainability. Strategies for Sustainability*. Springer, Cham. https://doi.org/10.1007/978-3-031-59622-3_2

Massaro, M., Secinaro, S., Dal Mas, F., Brescia, V., & Calandra, D. (2021). Industry 4.0 and circular economy: An exploratory analysis of academic and practitioners' perspectives. *Business Strategy and the Environment*, 30(2), 1213–1231. https://doi.org/10.1002/bse.2680

Mostaghel, R., & Chirumalla, K. (2021). Role of customers in circular business models. *Journal of Business Research*. https://doi.org/10.1016/j.jbusres.2020.12.053

Nikolaou, I. E., Jones, N., & Stefanakis, A. (2021). Circular economy and sustainability: The past, the present and the future directions. *Circular Economy and Sustainability*, 1, 1–20. https://doi.org/10.1007/s43615-021-00030-3

Pope, C., & Mays, N. (2020). The role of theory in qualitative research. *Qualitative Research in Health Care*. https://doi.org/10.1002/9781119410867.ch2

Purola, A., Santonen, T., Haapaniemi, H., Hirvikoski, T., & Nevmerzhitskaya, J. (2020). Living labs concepts and implementation plan for CIRC4Life-project. https://ec.europa.eu/research/participants/documents/downloadPublic?documentIds=080166e5c1ef0b12&appId=PPGMS

Salvador, R., Barros, M. V., Da Luz, L. M., Piekarski, C. M., & de Francisco, A. C. (2020). Circular business models: Current aspects that influence implementation and unaddressed subjects. *Journal of Cleaner Production*, 250, 119555. http://dx.doi.org/10.1016/j.jclepro.2019.119555

Salvador, R., Barros, M. V., Freire, F., Halog, A., Piekarski, C. M., & Antonio, C. (2021). Circular economy strategies on business modelling: Identifying the greatest influences. *Journal of Cleaner Production*, 299, 126918. https://doi.org/10.1016/j.jclepro.2021.126918

Salvador, R., Barros, M. V., Pieroni, M., Silva, D. A. L., Freire, F., & Antonio, C. (2023). Overarching business models for a circular bioeconomy: Systematising archetypes. *Sustainable Production and Consumption*, 43, 349–362. https://doi.org/10.1016/j.spc.2023.11.010

Sarkar, B., Ullah, M., & Sarkar, M. (2022). Environmental and economic sustainability through innovative green products by remanufacturing. *Journal of Cleaner Production*. https://doi.org/10.1016/j.jclepro.2021.129813

Sjödin, D., Parida, V., Jovanovic, M., & Visnjic, I. (2020). Value creation and value capture alignment in business model innovation: A process view on outcome-based business models. *Journal of Product Innovation Management*, 37(2), 158–183. https://doi.org/10.1111/jpim.12516

Staples, A. J., & Krumel Jr, T. P. (2023). The Paycheck Protection Program and small business performance: Evidence from craft breweries. *Small Business Economics*. https://doi.org/10.1007/s11187-022-00717-3

Subramani, R., Mustafa, M. A., Ghadir, G. K., Al-Tmimi, H. M., Alani, Z. K., Rusho, M. A., . . . Kumar, A. P. (2024). Exploring the use of biodegradable polymer materials in sustainable 3D printing. *Applied Chemical Engineering*, 3870–3870. https://doi.org/10.59429/ace.v7i2.3870

Tapaninaho, R., & Heikkinen, A. (2022). Value creation in circular economy business for sustainability: A stakeholder relationship perspective. *Business Strategy and the Environment*, 31(6), 2728–2740. https://doi.org/10.1002/bse.3002

Tuladhar, A., Iatridis, K., & Dimov, D. (2022). History and evolution of the circular economy and circular economy business models. *Circular Economy and Sustainability*. https://doi.org/10.1016/B978-0-12-819817-9.00031-4

Velenturf, A. P. M., & Purnell, P. (2021). Principles for a sustainable circular economy. *Sustainable Production and Consumption*. https://doi.org/10.1016/j.spc.2021.02.018

Verleye, K., De Keyser, A., Raassens, N., Alblas, A. A., Lit, F. C., & Huijben, J. C. (2024). Pushing forward the transition to a circular economy by adopting an actor engagement lens. *Journal of Service Research*, 27(1), 69–88. http://dx.doi.org/10.1177/10946705231175937

Zhang, Z., Qu, T., Zhao, K., Zhang, K., Zhang, Y., Guo, W., . . . Chen, Z. (2024). Enhancing trusted synchronization in open production logistics: A platform framework integrating blockchain and digital twin under social manufacturing. *Advanced Engineering Informatics*, 61, 102404. https://doi.org/10.1016/j.aei.2024.102404

10 Education and Workforce Development for the Circular Economy

Shuriya, K. Yogeshwaran, C. Ramesh, P. Balasubramani, K. Jagadeesh, and R. Gowri

10.1 INTRODUCTION

The circular economy is reinventing sustainability, breaking from the traditional take, make, and waste model. Instead, it promotes a more restorative system which supports the continual use of resources via recycling, reuse, repair, and refurbishment. It demonstrates a new approach in targeting the unsustainable linear consumption and production patterns that deplete natural resources, create too much waste, and damage work. In order for this shift to succeed and the promotion of circular economy approaches not just on paper but in practice, a workforce with specialist knowledge and skills is needed. And one of the great ways to achieve this is through education. This requires high levels of competence across transdisciplinary domains such as software processes, resource management, life cycle assessment, and sustainable design for people to be able to exploit the full potential circularity presents. Sustainable design—which involves the development of products and processes that have a low environmental impact—stands out as an effective way to stimulate rapid resource efficiency gains [1]. It also reflects notions such as eco-design and some other mechanisms to ensure material recovery in a purposeful, effective manner that is completely sustainable throughout the entire life cycle of a product from its conception through disposal. One of the fundamental elements of sustainable development is effective management of natural resources, such as responsible generation, consumption, and preservation throughout complete life cycle period. This means establishing policies to cut down on resource consumption and increase product lifetimes, including through recycling and reuse. This includes material flows, waste generation, and the occurrence of environmental burdens associated with resource extraction and processing [2]. Life cycle assessment (LCA) is a key tool for assessing the environmental impact of processes and products through their full life cycle. LCA provides decision-makers with key information on where to focus their attention in order to make changes and help achieve the goals of sustainability by evaluating a product's sustainability throughout its life: from production and extraction, through use-phase up until EoL. Given that digital technology and data analytics improve waste reduction, supply chain optimization, resource management, etc., software processes are a fundamental pillar of the circular economy. They enable the tracking

DOI: 10.1201/9781003538837-10

and application of materials in an innovative way, offering new opportunities for sharing/recycling resources or even contributing to establish fully circular goods [3]. That enable and facilitate new business models like PaaS service which spawns efficiency in resource use case by not buying thing, you only pay it for sharing or leasing instead. The skills required for people and organizations to fully embrace the principles of the circular economy, such as minimizing waste production, optimized resource flow, or lower environmental degradation. It is necessary that the key stakeholders such as the government sector, academia, civil society, and commercial organizations synergize one another in developing educational programs which are holistic to address circular economy [4]. One important area in which the ideas surrounding a circular economy can be incorporated is through academia, including within business studies, engineering, and environmental sciences. Many of these features allow students to obtain professional experiences with industry partners next door that will help support the education and what they learn by providing hands-on experience. Civil society also plays an important role in raising awareness of the circular economy and promoting sustainable consumption. We can all work together to create a more resource-efficient and sustainable future for economic resilience as well as the environment at large.

10.2 THE ROLE OF EDUCATION IN THE CIRCULAR ECONOMY

This change in focus calls for a highly skilled workforce to advance the circular economy and this is what education can offer. It provides people the basis of learning interdisciplinary subjects and skills that are in demand such as software processes, resource management, lifespan design evaluation, or sustainable design. Without these, at both the individual and organizational levels, there is a risk of resource flows not being minimized overly wasted, damaging significant environmental impacts.

10.2.1 Sustainable Design in Education

Located at the nexus of these benefits is sustainable design, which essentially refers to creating systems and products that fulfill societal, economic, and environmental demands—something central in constituent elements of circular economy (CE). We can see it demonstrated visually using Figure 10.1. Thus, integrating academic knowledge with practical skills is a key consideration in teaching sustainable design to students and educating them on how they can be actively involved in the CE [5]. For this type of full educational experience, curriculum development, project-based learning, and industry partnership are necessary.

10.2.1.1 Curriculum Development

(a) **Foundational Courses**

Foundational courses are necessary to teach when it comes to sustainable design. Eco design—makes goods with as little an environmental impact over the entire course of their life. Material selection courses teach students how to pick recyclable, durable, and eco-friendly materials. It is also necessary to teach students how to

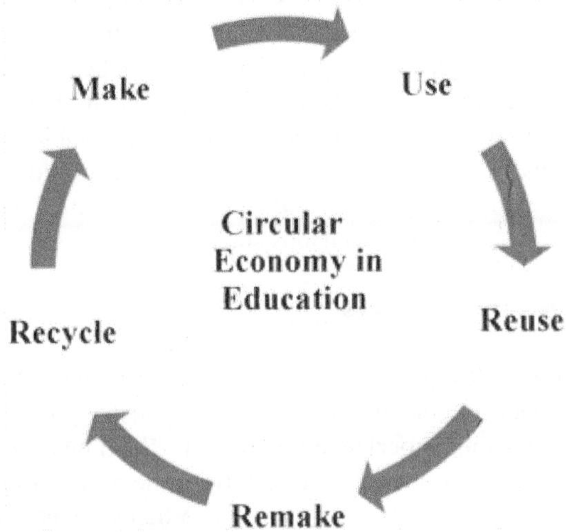

FIGURE 10.1 Sustainable design.

save energy and designing systems or products that consume less power, so it helps in reducing their carbon footprints [6]. Life cycle thinking seminars teach students to calculate the environmental impacts of each stage in a product's life—such as materials extraction, processing and manufacturing, and recycling or disposal. The knowledge and skills acquired by students who successfully complete these foundation courses prepare them to infuse sustainability into design methodologies.

(b) **Advanced Modules**

Advanced sustainable design further expands on the ideas and concepts covered in foundational courses, with more detailed and specialized subjects to study.

(c) **Design for Environment**

This might entail mitigating a product's impact from its creation until it is disposed of, as forwarded by designing for the environment [7]. To mimic natural cycles, the cradle-to-cradle approach aims at creating technologies that help perpetuate items and provide them with fitting process of recycling. Until then, biomimicry means creating solutions in a sustainable way by examining the patterns and mechanisms in nature. Upon successful completion of the advanced studios, graduates will be specialists in sustainable design on Bind view holder.

10.2.1.1.1 Collaboration across Departments

Teaching sustainable design cannot be taught in isolation for good outcomes. This has to be a coordinated effort and engineering, design, environmental science, and business all should participate. Students are allowed a more integrated insight into sustainable design: its context and complex consequences through an interdisciplinary integration of several academic disciplines. For instance, engineering and environmental science students can share the know-how regarding technical viability and ecological consequences while the design department teaches engineers how to value aesthetics functions [8]. This can also be done by business students who have to evaluate the market potential and financial feasibility of sustainable products. This collaborative effort helps students to develop a comprehensive understanding, making them well-equipped with the right tools and techniques in which they are required to solve sustainability-related issues at various stages of their practical settings.

(c) **Guest Lectures and Seminars**

Educational programs that regularly offer guest lectures and seminars by an array of academics, sustainability consultants, as well as industry professionals are necessary to enhance the learning experience for students. Such meetings can deal with a variety of themes about environmental design: from the newest tools and trends into real-world problems attended to by practitioners. When student get invited as guest lectures about their background, it helps them to know record of all information and experience beyond the conventional classroom settings [9, 10]. Guest lectures are important since it enables students to see how sustainable design concepts can be implemented in the real-world assort. Academicians also share their academic knowledge with real-time solutions through case studies, stories of professional experience in the field. It improves their learning and prepares them for future careers in sustainable design. Moreover, seminars are important for improving students' learning. These are interactive sessions where students directly have one-on-one interaction with subject matter experts from the industry. Your students can receive feedback on their own creations and ideas, chat with professional regarding those same thoughts, and ask themselves. Future designers strike this balance by existing halfway between theory and practice, to impart theoretical knowledge as well as practical abilities. Regularly inviting industry experts to be a part of their guest lectures and seminars makes institutions enhance the view they give students, beyond the ones that can only derive from textbooks or lectures. Students get exposed to the real-world expertise; it helps them understand better and also build a new imagination of thinking about how they can apply design principles into sustainable development.

10.2.2 Project-Based Learning

(a) **Design Challenges**

A big part of a sustainable design education is how to build stuff with the least harm to the environment. They design these curricular activities around essential sustainability principles such as durability, repairability, and recyclability. Dealing with such challenges endows the students with skills of critical thinking that involve both ability to recycle or dispose a product at its end-of-life, how durable is it, and does it readily repair if need be. This real-life approach reinforces theoretical knowledge while also developing creative and problem-solving skills when tackling genuine sustainability issues.

(b) **Life Cycle Assessment (LCA) Projects**

Life cycle assessment (LCA) is a method which allows the environmental impact of processes or products to be quantified over their entire life cycle. Every step of the way has to be considered if you really want to find areas where improvements can happen and this includes extraction, production, use once in the consumer's hands, and also disposal. Figure 10.2 illustrates fuzzy logic which is one such technique that combines empirical data and human judgment at each life cycle stage in order to provide a comprehensive coverage of environmental impacts. This method aids in providing decision-making over sustainability goals of different levels of complexities and uncertainties through scenario-based analyses.

Due to the uncertainty and ambiguous facts, fuzzy logic is often an efficient method for evaluating complex environmental implications [11–13]. There could develop fuzzy sets and membership functions to be able to measure environmental elements with respect to valued criteria. This adaptive context still considering changing conditions can allow for more sensitive and precise evaluation of environmental effects by all stakeholders. This ensures that decisions are evidence-based and responsive to changing contexts.

(c) **Multidisciplinary Teams**

Smart recruiters have found that combining the skills of top technical talent and business experts leads to results as close to real-world scenarios requiring expertise. This involves assembling students from various fields of study—engineering, design, environmental science and business—to work on projects in diverse topics ranging from industrial product to service design [14]. It is no different than the workplace; people of diverse types need to work together for formulating sustainable strategies. During this cooperative process, students earn crucial skills for communication as well as understanding each other strengths allowing them to have a comprehensive understanding of sustainable design.

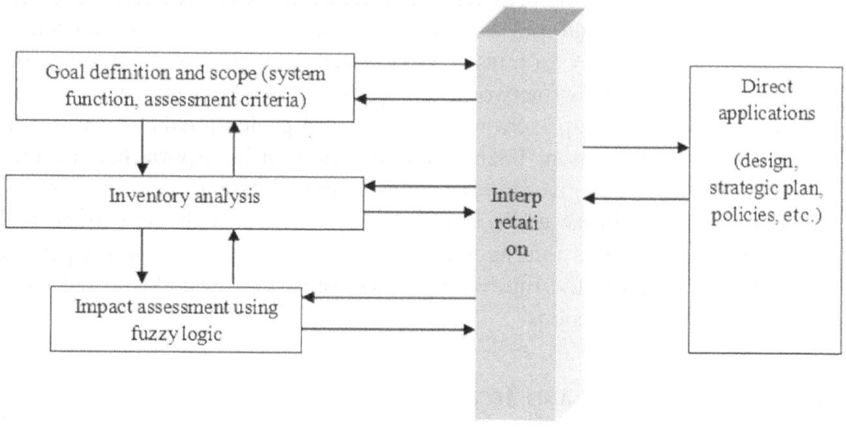

FIGURE 10.2 Fuzzy interpretation in LCA.

10.2.2.1 Mentorship Programs

Working with mentors from other disciplines or industries ensures that students receive more robust support and guidance on their projects, These mentors are there to provide useful feedback, give you insider perspective on how LifeChange classes run day to day and the course of a project. It helps them polish their concepts, breaks the misconceptions, and wards off a fear that helps to enhance self-confidence in students [15]. Mentors also help to bridge the gap between classroom learning and its practical application by giving student first-hand exposure of industry from professionals who have been there. Not only does this improve the capabilities of students, but it also draws a closer link between academia and business furthering the mentor´s career.

10.3 INDUSTRY COLLABORATION

10.3.1 Industry-Sponsored Projects

A smart approach to bridge the gap between academia and business is cooperation of companies with student projects. This means enables student to apply their learning in industry situation and solve real business problems for the companies [16]. Sponsored projects provide students with team dynamics skills, project coordination, and critical thinking, to name but a few. In addition, these relationships often become a platform for students to showcase their abilities before potential employers—possibly leading them with employment opportunities after graduation.

10.3.2 Internship Programs

There are many advantages to students when given the opportunity of being part of an internship program that specialize in sustainable design. It gives individuals an intimate look into the day-to-day operations of organizations working toward sustainability. Not only do internships allow students to track the real-world implications of class-based learning, but they also give them the opportunity to understand business operations and solving problems. Internships are the very first step to put all those theoretical concepts into practical learning and at least giving a knowledge of few crucial skills which play an integral part in your career growth. Besides, these internships come with a lot of networking opportunities across industries and give you guidance and mentorship from well-experienced professionals which can help you secure jobs postgraduation. There is a real benefit of incorporating sustainable design into internships, as it gives interns an in-depth understanding around concepts and practice of sustainability while also aiding businesses with their green ambitions. This not only benefits interns with a more enriched educational experience, and better job prospects in the future but allows companies to make strides toward their own sustainability mission.

10.3.3 Access to Tools and Technologies

Work design builds the business relationships that give students access to state-of-the-art design software, materials, and technologies for them to enter into today's market ready with what they need. This provides an opportunity for students to boost up their technical skills and make them familiar with the most modern industrial procedures. Examples include allowing students to meet sustainability criteria by enhancing their designs and rethinking ideas in a hands-on setting with state-of-the-art equipment.

10.3.4 Facility Tours and Workshops

This helps to bring the ideas of sustainable design from imagination into reality by showing students how they can be used in practical end-to-end environments through workshops and visits to facilities related with similar manufacturing. It influences students who visit these facilities and helps them learn how the business handles materials eco-friendly, as examples for other businesses one can talk about are a sustainable production practice and effective waste management approach. Experience with sustainable design principles practiced in real-world applications helps students appreciate and ultimately see the value of them. This knowledge will help them to improve upon their design efforts when they return back to lab or classroom.

10.4 IMPLEMENTATION STEPS

Step 1: Develop curriculum work with academics in related fields to develop robust and actionable curriculum for sustainable design. This will involve the selection of course content, determining requisite skills, and interdisciplinary education.

Step 2: Acquiring resources—secure funding and resources to engage with professionals in the industry, and introduce students to real-world project-based learning opportunities. They may have to find funds in seeking sponsorships or grants, as well as invest substantially on technology and tools.

Step 3: Faculty preparation—ensure that faculty has received and is prepared to deliver the appropriate training for teaching how sustainability can be brought into design (as cross-disciplinary subjects, etc.) along with managing multidisciplinary projects.

Step 4: Student engagement—promote the sustainable design program to create awareness among students about acquiring highly demandable skills in current labor marketing using flyers, platform commerce, info sessions, and social media feed.

Step 5: Ongoing improvements solicit feedback from industry partners and students on how can the program be improved next time by communicating frequently. This approach will allow you to start your project assignments and redesign course material according to the latest industry developments, as well as information related with sustainability research.

10.5 INTEGRATING FUZZY LOGIC IN CIRCULAR ECONOMY STRATEGIES

This chapter will creatively incorporate utilization of fuzzy logic for CE methods. This approach is particularly suited to the ambiguity and uncertainty inherent in circular systems [17]. This requires fuzzy logic to be transformed into educational systems that reduce the ambiguities generally occurring in CE for decisive and intelligent actions by professionals as well as students.

10.6 CHALLENGES IN IMPLEMENTING FUZZY LOGIC IN CE CONTEXTS

This is to resolve a few of the fuzzy logic application issues in circular economy (CE) which helps sustainability initiatives [18]. In this part, the main challenges are explored along with potential remedies that would make it less uncertain and more generally applicable.

10.6.1 Intricate Environmental Factors

The complexity of the environmental elements represents one of the most significant obstacles at which to use fuzzy logic, in CE situations. There are many things that can impact the environment in a variety of ways, most of them tangled up with multiple others and hard to get good numerical data on. Fuzzy logic offers a solution, using fuzzy sets and membership functions to describe ambiguous or uncertain data. Despite potential to be useful, however, it remains a major challenge in accurately describing these fuzzy sets and verifying its legitimacy across real situations. Fuzzy logic models have to rely on better quality data and expert knowledge in an effort not only to reduce the uncertainty, but also ensure more accurate environmental assessments.

10.6.2 Utilizing Current Systems for Integration

The main problem in this regard is the integration of fuzzy logic systems with existing CE frameworks and technology. The fuzzy logic is widely used in the construction of decision-making systems; therefore, this method can conflict with the current procedures and technologies of many companies [19]. The key challenge is developing interoperable solutions, which can seamlessly integrate fuzzy logic to the existing hardware and software components as well as operational procedures. In order to achieve this, cooperation between academic institutions and business, as well as technology companies is needed in the creation of adaptive systems that bring out effective operations without sacrificing environmental soundness.

10.6.3 Resources for Computation and Complexity

Fuzzy logic causes significant computational issues in the case of bigger systems and broader applications. Fuzzy logic methods are computationally expensive to perform for subset optimization in larger multiobjective decision-making, such as supply chain and smart grid applications [20]. Solving the problem is about being able to work out how best to deal with real-time analysis and massive data processing using research around distributed computing technologies, power of computation increases, and algorithms refined for maximum efficiency.

10.6.4 Standardization and Training

To gain the successful increasing use of applications in many circular economy (CE) subjects, it is necessary to work with standardized methods for fuzzy logic and make sure training are proper. Standardization is needed to ensure consistency in the conducting of environmental assessments and decision-making processes, which enhance interoperability and comparability of results among multiple industries. Second, fuzzy logic concepts will be widely accepted in industries when engineers and analysts have been trained properly [21]. Platforms for exchange of information, but also permanent training are necessary to increase our capabilities in basic and advanced applications of fuzzy logic in sustainable development. A potential solution to bundles of application-level scaling-problems that leverage fuzzy logic has thus far been elusive and will likely require a concerted effort between researchers, practitioners, and policy leaders. It is used to enhance decision processes in CE; however, despite its flexibility, certain issues must be addressed. Without the efficient optimization of computational resources, methodological standardization, superior data quality standards, and integration into existing systems coupled with numerous training access/availability to adopt this technology form at production level from day 1; else, industry would not be able to realize significant potential inputs in communal well-being through adoption path similar as fuzzy logic techniques. With these barriers taken down, fuzzy logic can be used creatively in the circular economy to promote resource efficiency and resilience as world changes so quickly now.

10.7 KEY COMPETENCIES FOR THE CIRCULAR ECONOMY

There are many key factors and strategies required for the software operations to implement circular economy [22]:

Software development methodologies: The circular economy would not be possible without agile and DevOps, as well as other iterative approaches. They are key to catalyzing the process of developing and enhancing circular economy solutions. Rapid prototyping and iterative improvement methods that prioritize short-range planning then allow companies to adapt quickly in the face of fluctuating demand or feedback. As such, they are integral to the continued support for operational efficiency and sustainable design principles in the circular economy, respectively.

Data analysis and management: Big data and advanced analytics are essential to monitoring the flow of resources, forecasting future trends, and implementing evidence-based plans in a circular economy. Software tools can collect data on material input, energy consumption, and waste generation at scale which can be screened and interpreted in a comprehensive fashion. This allows you to uncover inefficiencies and opportunities for more effective resource usage with less environmental footprint.

Simulation and modeling tools: For turning systems from the physical world into virtual representations, software modeling and simulation methodologies are indispensable. They have the ability to run scenario analysis, prediction, and optimization without needing any physical testing for a deeper understanding of what you are working with. Thus, businesses might evaluate the environmental and resource impacts of different endeavors earlier than their implementation. Moreover, such technologies allow researchers and engineers to experiment with their ideas without the cost of trial-and-error approaches.

Digital twins: In the circular economy, this necessity is all the more vital and digital twins support real-time performance monitoring that helps in consistent optimization. They use sensor data and other information to provide insights into the effectiveness and behavior of real systems or processes. They are crucial for the progression of sustainability in the circular economy as they allow for predictive maintenance and process efficiency, which minimizes resource wastage through real-time data-based optimization.

Adjoining these software operations to the circular economy offers numerous perks, like better judgment skills and capability of using resources in a smarter manner while even more inducing industries to draft sustainability measures. Through digital twins, simulation tools and agile methodologies coupled with the analysis capacity of large data industry organizations can modify processes to harmonize best practices circularity. That in turn is what makes it sustainable over the long term and reduces waste on your local landfill, etc.

10.8 CASE STUDIES

To illustrate the potential value of CE education, this chapter also provides several case examples from different industries where Design and CE are working used together:

10.8.1 Manufacturing

The application of a circular economy (CE) model is widely used in the manufacturing industry for waste prevention and resource efficiency on a build-to-order basis.

TABLE 10.1
Manufacturing Industry: Case Study Perspective

Company ID	Size	Role	Industry Domain
A	MNE	Global	24 Manufacture of fabricated material
B	SME	Subcontractor	19 Manufacture of fabricated metal
C	MME	National	15 manufacturing food products
D	MME	National	15 textile manufacture
E	SME	Global	13 manufacturing lathe machines
F	SME	Global	
G	MME	Global	14 Manufacture of electrical equipment
H	SME	Global	21 waste collection and recycling
I	Nonprofit organization	Subcontractor	9 manufacture textile

This approach often includes expensive, time-consuming efforts to train staff on how to achieve these goals. This perspective is illustrated for the industrial sector in Table 10.1 case study view, where possible references to other contributions are provided as well.

CASE STUDY: MANUFACTURING COMPANY

Objective: Educating workers to develop a circular economy strategy.

Background: A consumer electronics manufacturing company has been awakened to the need for sustainability and this resulted in their choosing a circular economy route to improve resource productivity and reduce environment degradation.

Implementation Steps

Step 1: Employee Training Program Development

Topics in the training program will include practices for sustainable production, waste minimization, and effective resource management with a special focus on the circular economy concept. The program aimed to create awareness about the importance of sustainability and understand how they can help achieve goals set for a circular economy among everybody in an organization from entry-level employees up to management.

Step 2: Content and Modules

- *Exploring the circular economy:* This introduced workers to the concept of a circular economy, highlighting how it contrasts with traditional linear manufacturing that follows take–make–dispose practices.
- *Intermittent resource utilization:* Training workshops were focused on methods for optimizing the use of resources in manufacturing processes, which included using energy efficient technologies and best practices, improving raw material utilization, and minimizing water usage.
- *Waste minimization:* Staff were trained in the reduction of waste methods—wood recycling and reusing techniques. These practices are crucial to the growth of closed-loop manufacturing.
- *Case work:* Staff were trained on real-life case of different sectors where circular approaches have been proven to be successful.

Step 3: Handson Workshops and Simulations

The training program incorporated practical workshops that enabled staff to practice their newly gained skills in manufacturing simulations. Through these interactive simulations, employees were able to research various scenarios and see how their choices impacted waste output as well as resource use. The workshops were in practical hands-on style and the staff members practiced on manufacturing simulations. The interactive simulation allowed employees to explore what-if scenarios and gain an appreciation for how their decisions impacted trash generation and resource consumption.

Step 4: Monitoring and Feedback

The company has also implemented mechanisms to assess the impact of its training programmed periodically. This will cover reviewing waste minimization and resource efficiency–related performance indicators, as well as conducting feedback sessions with stakeholders. Based on this data, processes for ongoing improvements are put into place to increase the efficacy and operations of a program.

Outcomes

- *Better resource utilization:* There were marked improvements in resource efficiency metrics after organization implemented strategies taught in the training. This is exemplified by the reduction in material used and a decrease in energy consumption per unit produced.
- *Waste production was lower:* The more the employees and even customers knew how to recognize when waste had chances for recycling, or repurpose in one of those production processes.
- *Employee engagement strategy:* The training was organized to generate an innovative and sustainable spirit among the organizing staff so that they come up with various ways of improving environmental outcomes.

10.8.2 Construction

A curriculum that created innovative, environment-friendly building projects by integrating sustainable construction concepts.

CASE STUDY: EDUCATIONAL INITIATIVE IN CONSTRUCTION

Objective: Integrating circular economy (CE) principles into construction education.

Background: Sustainability in building performance-driven facility managers. Students wishing to enter the workforce with a basic understanding of sustainable building practices were seen as equally valuable, an opinion shared by the administration at one school. They restructured the curriculum to add CE concepts so as to promote creativity in sustainable building initiatives.

Implementation Steps

Step 1: Curriculum Integration

Over the years, changes were made to its programs to allow other departments within the college (e.g., construction management, architecture; engineering and environmental studies) opportunities for inclusion of CE concepts into a variety of courses, so it recently revised its curriculum. It also introduced more modules—life cycle evaluation tools, circular design fundamentals, and sustainable building materials.

Step 2: Project-Based Learning

The training curriculum combined project-based learning modules built on CE principles with experiential learning techniques. Across the board, students were able to design creative sustainable building solutions with a consideration on resource management, waste reduction, and use of renewable resources.

Step 3: Interdisciplinary Collaboration

The importance of having a team comprised of various disciplines such as engineering, architecture and environmental sciences was stressed to further advance the comprehension on sustainable construction. So interdisciplinary teams were formed, as they would be in real life for green building design professionals working together on projects.

Step 4: Guest Lectures and Workshops

Industry professionals specialized in sustainable building and circular economy procedures were invited to take guest lectures, workshops for the students as part of this initiative. In order to motivate students, they used case studies, talked about their own journey experiences, and mentioned about what can be done best. This delivered practical experience beyond the regular classroom syllabus.

Step 5: Field Trips and Site Visits

Field visits to eco-friendly building material manufacturing plants and sustainable construction sites. It was through these experiences that they finally came to understand why sustainable architectural ideas are so hard to implement and many examples of their applications. In addition, such revisits provide students vital thoughts that can relate to their own practice.

Outcomes

The curriculum has been reframed to incorporate circular economy (CE) principles that have given rise to innovative sustainable building projects. Among many examples, such initiatives have demonstrated how to keep life cycle analysis simple while comprehensive in tackling complex waste management challenges and material recycling as well as energy conservation. It has proven to be invaluable as it gives students hands-on experience solving unclear sustainability issues and turns their already-obtained theoretical knowledge into real-life applications. This hands-on mindset puts students in a position to propel sustainability forward proactively. As a result, many in the industry have taken note of their efforts to produce students who are well-versed in CE principles. Employers looking to implement sustainable construction methods actively seek out these graduates for their skills and knowledge. In the end, not just promoting innovation in construction industry through its embodied approach, but it also intents on producing responsible sustainable construction professionals by embedding CE ethos within its curriculum. Part of an effort is to teach the construction industry's future leaders about sustainability by both addressing current environmental issues and making it part of their planning process. Aside from just short-term gains, it sets a benchmark for the development of an environmentally responsible industry well-served to society as a whole.

10.8.3 Retail

A collaboration between a retail company and a higher education institution aims to develop a consumer electronics supply chain management training program.

CASE STUDY: COLLABORATION ON CIRCULAR ECONOMY TRAINING IN RETAIL SUPPLY CHAIN MANAGEMENT

Objective: Developing a circular economy (CE) training program for supply chain management in retail

Background: After seeing potential for streamlined sustainability in its supply chain, a retail company worked with a university to implement customized training program. The collaboration is about bringing circular economy knowledge into supply chains, aiming to drive the sustainability transition in retail movements.

Implementation Steps

Step 1: Needs Assessment and Curriculum Development

To identify potential improvements in the integration of CE principles into supply chain management, a full needs assessment was conducted with collaboration between the university and retail company. Consequently, there is now a curriculum that focuses on vital topics such as reverse logistics best practices, finished goods inventory strategies and policies (foresight planning), sustainable procurement solutions for raw materials/manufactured components, and effective distribution methods.

Step 2: Expert-Led Course Delivery

In these courses, students learned from professors at the university and CE managers who are seasoned pros with hands-on supply chain experience. To provide students the theoretical knowledge and practical skills of managing supply chains in retail, these sessions were divided into lectures, case studies, and workshops.

Step 3: Industry-Driven Projects

The program provides students access to industry-based projects in partnership with a retail business. Most of these projects will center on instilling sustainable practice in the corporate supply chain, and activities such as advanced packaging waste-minimization technologies, efficient transport-routing systems that result in lower carbon emissions, or other broader focused reverse logistics initiatives (for returnable management) are more-or-less to be expected.

Step 4: Internship Opportunities

Internship opportunities in supply chain for students as part of a program run by the retail corporation's supply chain branch. This means that they could apply the things learned from CE knowledge to work with subject matter experts for managing their sustainability projects. The experience allowed interns to follow the changes and see how they were improving or worsening efficiency in their supply chain.

Step 5: Evaluation and Continuous Improvement

Regular feedback sessions were employed to evaluate the effectiveness of the training program. These assessments were then employed to revise the case studies, secure more recent supply chain management, and CE developments along with enhancing course material.

Outcomes

Having implemented a training program, the supply chain of the major retailer has been verified as significantly more sustainable. This initiative has focused on means like resource efficiency, waste minimization, and lifespan expansion to promote sustainable practices throughout the supply chain. As an example, the partnership between a university and retail company generated creative solutions and best practices in accordance with circular economy (CE) basics through fostering knowledge transfer from academia to business. Aside from solving environmental problems, an initiative like this shows how cooperation can help to ensure more workable business sustainability plans. Graduates from this program have bagged CE experience helping them with job placement against other graduates and got bigger roles in their retail company. This initiative has further validated the strong commitment of this retail company to sustainable supply chain management. Through their cooperation, both organizations have made significant contributions to establishing a more sustainable future for retail by creating a CE training program specifically designed with supply chain management in mind. This is the kind of public–private partnership that can make real progress against pollution in every sector. With this example, these are great ways in which teams can work together to benefit of both the organizations and our society.[1]

10.8.4 Software Development

One technology company which integrated the CE ideas in its software development process then devised to implement such tools for other businesses so that they may reduce their waste as well as save on resources.

CASE STUDY: INTEGRATION OF CIRCULAR ECONOMY PRINCIPLES IN SOFTWARE DEVELOPMENT

Objective: Developing Software Tools to Optimize Resource Use and Minimize Waste

A software development agency recognized an opportunity to innovate in sustainability by integrating circular economy principles into its company; As a company, we wanted to enable other companies or businesses related to any sector who are seeking digital solutions for waste reduction and better resource utilization.

Implementation Steps

Step 1: Integration of CE Principles in Software Development:

As an example, a software company adopted CE principles (e.g., tool chain to reduce resource consumption and waste) into their product development process. It adopts design for sustainability and sustainable development practices with the help of Agile, DevOps to take care of some environmental aspects. These pseudo-methods focus on the incremental improvements, and make software development life cycle sustainable.

Step 2: Tool Development for Resource Optimization

The company came out with specialized software tools to assist the firms in managing their resources flow through the operations. By mashing up data from cutting-edge tools to measure waste generation, material use, and energy consumption across board geographies, Smart World Loc sentimentally operates on the chief aim of unearthing regions as well as places that are not functioning at peak efficacies.

Step 3: Life Cycle Assessment and Optimization

Specifically designed for conducting life cycle evaluations (LCAs) of digital goods and services, the software tools complement each other well. These life cycle assessments allowed us to understand the environmental impacts at each stage of a smartphone (development, deployment, continuous use phase, and end-of-life management) which is incredibly important! This made businesses able to take informed decisions that would reduce negative environmental impact and promote sustainability.

Step 4: Collaboration with Business Clients

To better understand clients' unique sustainability needs, I worked directly with technology companies, including airlines and global manufacturers, focusing on time management. They offered bespoke services such as integration and customization to adapt software solutions in the context-specific operational environment and sustainability goals of each client.

Step 5: Training and Support

The company also supported customers in their efficient utilization of CE-focused software instruments via comprehensive training events and continuous assistance. Knowledge of software functionality, the capacity to analyze sustainability KPIs, and the ability to apply proven techniques all were topics discussed during training to make the participants' integration of sustainability in their business procedures as seamless as possible.

Outcomes

The IT start-up has created a set of software tools that will revolutionize how companies view sustainability. These technologies can help businesses eliminate waste, minimize resource use, and improve overall sustainability performance. The software has been designed with the circular economy in mind, leading to a huge reduction in resource and waste production by clients. Aside from the environmental benefits, this creates a significant edge for companies compared to their respective marketplaces. As a result, the IT company is able to attract customers who are committed toward achieving environmental sustainability targets, establishing itself as a front-runner of sustainable technology solutions. Incorporating CE principles into software development is a way to support this transition from businesses (and society as large) toward more sustainable practices. This case study provides a great example for the impact of technology on environmental outcomes as well as international efforts toward realizing circular economy.

10.9 CONCLUSION

Training and capacity-building are crucial to the growth of a circular economy. We can facilitate a group of professionals who know what they are talking about, and thus better advocate for sustainable practices. In order to improve decision-making and treat the circular systems in all its complexity, it is essential that contemporary approaches such as fuzzy logic and digitized tools are integrated into educational models. By addressing the challenges of curriculum design and faculty training, we can more easily scale up a strong educational infrastructure for the circular economy that promises to help create communities that are far less carbon polluting.

NOTE

1. http://www.gmdu.net/corp–335312.html

REFERENCES

1. Corvellec, Hervé, Alison F. Stowell, and Nils Johansson. "Critiques of the circular economy." *Journal of Industrial Ecology* 26, no. 2 (2022): 421–432.
2. Mont, Oksana, and Raimund Bleischwitz. "Sustainable consumption and resource management in the light of life cycle thinking." *European Environment* 17, no. 1 (2007): 59–76.
3. El Alaoui, M. "A fuzzy multiplicative performance indicator to measure Circular Economy efficiency." *International Journal of Mathematical, Engineering and Management Sciences* 5, no. 6 (2020): 11–18.
4. Carruth, D. W. "Virtual reality for education and workforce training." In *2017 15th International Conference on Emerging eLearning Technologies and Applications (ICETA)* (p. 16). IEEE, October 2017.
5. Kagan, Sharon Lynn, Kathleen C. Tarrant, and Kristie Kauerz. *The Early Care and Education Teaching Workforce at the Fulcrum: An Agenda for Reform.* Teachers College Press, 2008.
6. Geissdoerfer, Martin, Paulo Savaget, Nancy M. P. Bocken, and Erik Jan Hultink. "The circular economy–a new sustainability paradigm?" *Journal of Cleaner Production* 143 (2017): 757–768.
7. Wang, Jingjing, Nur Azlina Mohd Mokmin, and Shuang Ji. "enhancing higher education art students' learning experience through virtual reality: a comprehensive literature review of product design courses." *Interactive Learning Environments* 32, no. 10 (2024): 7399–7415.
8. MacArthur, Ellen. "Towards the circular economy." *Journal of Industrial Ecology* 2, no. 1 (2013): 23–44.
9. Murray, Alan, Keith Skene, and Kathryn Haynes. "The circular economy: An interdisciplinary exploration of the concept and application in a global context." *Journal of Business Ethics* 140 (2017): 369–380.
10. Jacquemin, Leslie, Pierre-Yves Pontalier, and Caroline Sablayrolles. "Life cycle assessment (LCA) applied to the process industry: A review." *The International Journal of Life Cycle Assessment* 17 (2012): 1028–1041.
11. Roy, Poritosh, Daisuke Nei, Takahiro Orikasa, Qingyi Xu, Hiroshi Okadome, Nobutaka Nakamura, and Takeo Shiina. "A review of life cycle assessment (LCA) on some food products." *Journal of Food Engineering* 90, no. 1 (2009): 110.

12. Tillman, Anne-Marie. “Significance of decision making for LCA methodology.” *Environmental Impact Assessment Review* 20, no. 1 (2000): 113–123.
13. Balleisen, Edward J., Lindsay Howes, and Erik Wibbels. “The impact of applied project-based learning on undergraduate student development.” *Higher Education* 87, no. 4 (2024): 1141–1156.
14. Wu, Jing, and Ayodele T. Olagunju. “Mentorship in medical education: Reflections on the importance of both unofficial and official mentorship programs.” *BMC Medical Education* 24, no. 1 (2024): 1233.
15. Gangrade, Nikita, Christy Samuels, Huda Attar, Allison Schultz, Natacha Nana, Emily Ye, and William M. Lambert. “Mentorship interventions in postgraduate medical and STEM settings: A scoping review.” *CBE—Life Sciences Education* 23, no. 3 (2024): ar33.
16. Maspul, Khairil Anwar. “Enhancing project-based learning in STEM education with integrated technology and coding.” *Journal of Intelligent Systems and Information Technology* 1, no. 1 (2024): 16–24.
17. Hoke, Amanda M., Krista Greenawalt, Emily VanDyke, Allison Zearfaus, and Deepa L. Sekhar. “Evaluation of the inaugural Pennsylvania school nurse mentorship program.” The Journal of School Nursing (2024): 10598405241301995.
18. Nikolakis, Nikolaos, Paolo Catti, Alexis Chaloulos, Wilhelm van de Kamp, Mildred Puerto Coy, and Kosmas Alexopoulos. “A methodology to assess Circular Economy strategies for sustainable manufacturing using process ecoefficiency.” *Journal of Cleaner Production* 445 (2024): 141–289.
19. Saari, Ulla A., Svenja Damberg, Matthias Schneider, Leena Aarikka-Stenroos, Cornelius Herstatt, Minna Lanz, and Christian M. Ringle. “Capabilities for Circular Economy innovation: Factors leading to product/service innovations in the construction and manufacturing industries.” *Journal of Cleaner Production* 434 (2024): 140295.
20. Patwa, Nitin, A. Seetharaman, Aashin Arora, Rishik Agrawal, and Harshal Mandalia. “Circular economy: Bridging the gap in sustainable manufacturing.” *The Journal of Developing Areas* 55, no. 1 (2021).
21. Chen, Lijuan, Ditao Duan, Arunodaya Raj Mishra, and Melfi Alrasheedi. “Sustainable thirdparty reverse logistics provider selection to promote Circular Economy using new uncertain intervalvalued intuitionistic fuzzy projection model.” *Journal of Enterprise Information Management* 35, no. 4–5 (2022): 955–987.
22. Renfors, Sanna M. “Education for the circular economy in higher education: an overview of the current state.” *International Journal of Sustainability in Higher Education* 25, no. 9 (2024): 111–127.

11 Case Studies and Practical Examples in the Field of Circular Economy

Aditi Bhardwaj, Pankaj Sharma, and Vinay Jain

11.1 INTRODUCTION

Among the largest alters happening in our society today is the embrace of circular economy. Overconsumption and resource scarcity are issues that require creative solutions that foster innovation and the growth of new businesses. This entails switching to a circular economy that is predicated on closing material cycles, from a linear model, which concentrates on utilizing raw materials to manufacture a product and then discarding it after usage [1–3]. Reducing, reusing, recycling, and healing materials in the processes of production, distribution, and consumption while concurrently creating economic prosperity, social equity, and environmental quality for the benefit of present and future generations are the strategic focus of the shift to a circular economy [4]. Leading the charge in advancing the circular economy is the European Union. The "Circular Economy Package," introduced by the European Commission in 2014, might be considered as the apex of prior EU policies on certain facets to the circular economy, enhanced by fresh perspectives [5, 6]. Additionally, specific prohibitions on plastics that are only used once were implemented in 2018 [7]. Furthermore, certain member states have taken unique actions to promote a circular economy [8]. Despite the benefits provided by these incentives, the EU is taking a long time to adopt the circular economy. As per Ref. [9], the primary cause of this is not technology constraints but rather cultural barriers like reluctance on the part of the company culture and restricted consumer interest. Notwithstanding these limitations, an increasing amount of projects have lately been established in a number of European towns and areas, including the Netherlands. Their objective is to provide motivational models for implementing measures that promote a circular economy, thereby enhancing the welfare of society, the environment, and the economy [10, 11].

11.1.1 Overview of Circular Economy

The ideas of the CE (circular economy) have been completely embraced by a number of countries throughout the world, replacing their previous waste management techniques and regulations [12, 13]. Most of these nations are similar in that they

 DOI: 10.1201/9781003538837-11

have developed economies, cutting-edge technology, strong political resolve, reputable public administration, and cutting-edge public infrastructure. This has brought up the question of whether low- and middle-income nations can embrace the same circular economy in all of its forms—theory, practice, and policy. Lack of information and ignorance of the implementation procedures are further problems that contribute to the absence of CE in the continent [14, 15]. While CE acceptance has been widely documented as having practical benefits at the national, subnational, and local administration levels as well as in the business community in the countries where it is currently implemented, in developing countries—aside from G20 nations like China—it is still an emerging idea [16–18]. A new ecological paradigm called the circular economy has recently surfaced to address resource shortages and global warming [19]. CE has been conceived as a replacement to the present sequential economy, which is predicated on a "take–make–dispose" paradigm, by disentangling economic development from resource exploitation [20]. Implementing closed-loop supply networks built on practices like recycling, remanufacturing, and reusing are examples of CE techniques. Other tactics include selling the purpose rather than the good alone and employing collaborative and sharing business models [21, 22]. The capacity to use resources in several cycles and reduce waste by delaying, shutting, or narrowing the loop is what gives a CE its value rather than boosting sales and material flows [23, 24]. However, the application of CE in manufacturing enterprises is still in its infancy [25–27], and in order to restructure sectors as well as supply chains to comply with CE, a number of obstacles must be addressed [22].

11.1.2 Importance of Practical Implementation

To achieve "real-life action of circular economy," one must, nonetheless, venture outside the realm of theoretical thought. The practical demonstration of the feasibility and advantages of a circular economy is provided by the implementation of circular economy tactics in real-world environments. Organizations that use circular systems, for instance, reduce waste, increase resource productivity, and pay less for production and procurement. They become more competitive as a result, and it also helps them achieve the sustainability goals. Innovative ideas are also sparked by practical application, as businesses come up with novel ways to dispose of, remanufacture, or reuse certain resources. Furthermore, it fosters synergy by motivating enterprises, consumers, and the government to create a coordinated environment. In this regard, case studies of the manufacturing programs of Renault and the efforts of the Ellen MacArthur Foundation show how circular economy practices benefit both the economy and the environment, or how the theory is turning into a reality [28, 29].

11.1.3 Role of Fuzzy Logic in Circular Economy

Fuzzy logic is an innovative instrument for advancing the circular economy since uncertainty and ambiguity are prevalent characteristics of many modern challenges. Fuzzy reasoning relies on logical values ranging from 0 to 1, which suggest a degree of truth instead of an either–or situation, in contrast to traditional two-valued logic.

Consequently, fuzzy logic can help with improved resource usage, waste management, and action sustainability within the context of a circular economy. Fuzzy logic models, for instance, are capable of evaluating the viability of materials by gauging several aspects of the material and its surroundings, such as the material's capacity for recycling, the degree of contamination it contains, or the material's economic viability in terms of recycling [30]. This is especially helpful in urban waste management systems since there are a lot of variables and unforeseen circumstances [31]. Important factors include the use of resources needed, effects incurred on the environment, and the amount of costs incurred. It makes it practical to build new stronger supply chain networks which are the fundamental for the sustainable use of resources [32]. In addition, the analysis of stakeholders' attitudes toward industrial waste is facilitated by fuzzy logic, which accommodates various specifics and imprecise data. This evaluation [33] considers the nature of the study's objective elements, analyzing at least four quantitative values that correlate with both the type and nature of the evaluation proposed in the referenced sources [33]. Fuzzy logic is included into circular economy activities to facilitate the development of good strategies for transforming the economy into a high-resilience, sustainable peripheral [34].

11.2 CASE STUDIES IN URBAN PLANNING AND INFRASTRUCTURE

These days, research on CE and its potential to interact with and contribute to sustainable development concerns is being conducted across a number of disciplines, from finance to urban planning [4]. Though the idea of CE has been widely accepted, no clear-cut definition or consensus has yet been reached [35]. The concept of CE is still developing [36–38], rather nebulous and unclear [39], and its full potential "still needs to be unlocked."

One of the primary goals of CE strategies is addressing the built environment, considering its significant impacts to a number of environmental challenges [40]. Nevertheless, there is still a dearth of scholarly research on the topic [41, 42], and the notion has only just begun to be used sporadically in real-world settings [41, 43, 44]. The CE is mostly considered to refer to trash management and recycling [45], and there has not been much research done on the possible consequences of applying it to urban areas [46]. Furthermore, there is not much agreement on how to approach and handle this idea in the building industry, and the skills and resources needed to implement it are still in the early stages of development [47].

11.2.1 Amsterdam's Circular Economy Strategy

The idea of the "circular economy" has gained a lot of traction because to the advocacy of business professionals and organizations like business foundations, business associations, and business consultancies [48, 49], as well as the growing support of national and international policymakers. For example, the World Economic Forum's [50] has issued studies on the circular economy, which today influences EU policy [51] as well as local and city policy [52].

The purpose of our exploratory case study is to get a deeper understanding of the governance structures that both support and are supported by municipal imaginaries of the circular economy.

Our goal is to better understand how public authorities who are positioned to promote local urban transformations and inspire circular urban ideas globally are co-producing, translating, and making information about circular economy visible. We choose to concentrate our research on European cities, particularly those that have identified itself as being at the vanguard of the European adoption of the sustainable economy within urban environments, due to our close cultural and geographical linkages. We chose three cities in particular because they are more likely to have an impact on information collaboration on the urban circular economy at both regional and global scales due to their well-known status as worldwide metropolises and their extensive worldwide networks [53].

London, Paris, and Amsterdam are these cities. There are two key numbers that separate these three cities:

(1) Greater London has a population of 8.6 million, Greater Paris has a population of 6.9 million, and Amsterdam has 840,000 people.
(2) The economic structure of the three cities is as follows: Paris's economy is characterized by a strong tertiary focus, but it also has important companies in the fields of aeronautics/aerospace, data technology, interactions, and biotechnology.
(3) Amsterdam is the economic and operations capital of the Netherlands and the fourth largest port in Europe.

The long-term goal of "Gemeente Amsterdam," also known as the City of Amsterdam or the Amsterdam Municipality, is to recover resources by separating 65% of waste from homes by 2020 and to create new methods of production, distribution, and consumption through innovation, research, and new practices. "The ability to recognize and carry out sustainable initiatives at the city level will lead to the creation of jobs, a revitalized or new industry, and increased competitiveness in global markets," the prediction goes. Many of the environmental, economic, and geopolitical problems that cities throughout the world are facing can be solved by the circular economy [54].

The municipality commissioned a study of the potential, which was conducted using the "Circular City Scan approach." The analysis determined where and how circularity should be implemented, as well as what effects may be expected in terms of jobs, ecological advantages, and additional value. According to the findings, Amsterdam "has the potential to realize economic growth and stimulate employment opportunities while, at the same time, greatly reducing emissions of greenhouse gases and material consumption." The City of Amsterdam accounts for 47 billion of the 106 billion euros in yearly economic activity in the Amsterdam metropolitan area [54]. The municipal government of Amsterdam has dedicated itself to implementing the following six guiding principles in order to accomplish these objectives [55]:

1. There is no waste since every substance ends up in an endless biological or technological cycle.
2. All energy will come from sustainable sources.

3. New financial or nonfinancial advantages will result from the use of natural resources.
4. Supply chains and modular, adaptable product design will facilitate system flexibility.
5. In order to shift from owning to using services, new economic models for production, distribution, and consumption will be created.
6. The services provided by ecosystems and the restoration of "natural capital" will be aided by human activity. The City of Amsterdam has pledged to become a "sharing city" in addition to being a "circular city," which entails "encouraging activities in the economy of sharing that promote innovation, social inclusivity, entrepreneurship, and sustainability" [53]. The city describes the sharing and circular economy initiatives as separate but concurrent activities.

On the one hand, sharing is mentioned as a tactic in the circular economy agenda, but it is only mentioned in passing and in connection to the usage of certain products (like vehicles), not as a necessary component of a systemic transformation. However, the circular economy is not included in the sharing economy, a strategy that the City of Amsterdam released in 2016. "The sharing economy is not an issue of prohibiting or allowing, but of observing and seizing possibilities where possible" [53] characterizes the sharing activities as the output of a business-driven online system, independent of intervention from the governmental authorities.

11.2.2 Urban Mining and Material Recovery

"Closing the loop," "securing resources," and "joining the circular economy" are becoming trendy. Phrases like the circular economy, ecological design, ecological impact printing, and zero waste, as well as processes like recycling, recuperation of resources, urban and garbage mining, waste minimization, and material recovery, are being utilized more often by politicians, businesspeople, and industry players. This shows that society has finally caught up with the garbage and resource administration community in realizing the need of recovering and using resources found in trash to the fullest extent feasible. This momentous occasion is due to a number of factors, such as growing concern over the world's rising nonrenewable resource consumption, steadily declining primary raw material supplies, dwindling space for waste disposal, the need to reduce waste production in terms of both volume and quantity, the need to control pollution of the environment from waste treatment emissions, shifting public perceptions of waste management, etc. [56]. The process of recovering chemicals and components from any type of anthropogenic stock, such as buildings, infrastructure, industries, goods (both in and out of use), environmental media absorbing anthropogenic emissions, etc., is referred to as "urban mining" and goes beyond landfill mining [57, 58]. The concentrations of elements in the stored materials are frequently equal to or greater than those in wild stocks, making them potentially a substantial supply of resources. Regarding natural ores, it is necessary to extract and treat anthropogenic stocks in order to provide economic gain. Due to these factors, electrical and electronic components (WEEE), which include comparatively large quantities of pricey metals and elements from rare earths, were the initial target of urban mining. Energy that may be produced via managing and treating trash in addition to recycling materials is referred to as resource recovery. The goal of materials recycling is to turn some wastes into resources that may be utilized to create new goods. Waste flows that are often categorized as belonging to material recycling schemes include packaging waste (plastics, paper, cans, glass), putrescible, bottom ash, sewage, exhausted oils, scrap tires, WEEE (or e-waste), end-of-life vehicles, etc. Following processing, which need not involve an extraction procedure, the recovered components are reintroduced into production cycles [56]. Although the phrase "urban mining" should primarily refer to the extraction of anthropogenic stocks based on its etymology, it is currently used to describe nearly any type of material recycling. A definition that differentiates between anthropogenic and natural resources' stock and flow may be required to provide language clarity [58]. The yearly inventories of materials stored in infrastructure, groundwater reservoirs, residential and commercial buildings, infrastructure, and scrap goods may not change much over time. Nonetheless, yearly material flows can vary significantly from one year to the next, contingent on factors like fashion, technological advancements, and the state of the economy. However, secondary raw materials are created from both anthropogenic stocks and flow resources (refer to Figure 11.1).

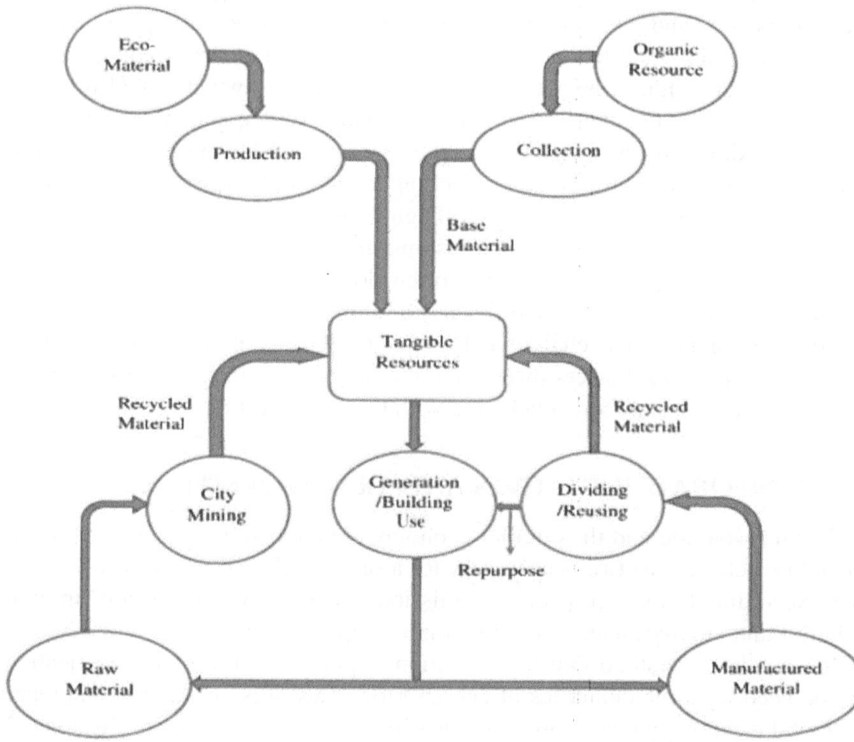

FIGURE 11.1 The characteristics of the material fluxes between various material resource sources.

11.2.3 Digital Platforms for Tracking Material Flows

For both financial and environmental reasons, the circular economy represents a fundamental advancement in material economics. Approximately one-third of the emissions of greenhouse gases (GHG) in the world are caused by industry. In a scenario that complies with the Paris Agreement, materials make up around 75% of the co-budget allotted to industry at the European level [59]. Reduced extraction, longer product and material lifetimes, and waste prevention all are important components of CE measures that may help lower greenhouse gas emissions and protect the environment. Reusing and recycling materials also lowers costs and has a positive environmental impact. However, the circularity gap that is presently stressing the global economy and environment indicates that CE is not happening just yet. Two years ago, the global economy was 9.1% circular, but only 8.6% of it is now [60]. There are high hopes for a technologically enabled CE [61–63]: technological advances have been recognized as essential facilitators of CE business models [64], and they also present a significant chance to boost business productivity and resilience while bolstering sustainability and more effective industrial systems. Enabling practices such as proactive upkeep, refurbishment, or recycling requires the collection and retention of information about the quantity, quality, and material content of products. Only then will the enormous opportunities presented by the CE be fully realized through the coordination of both information and material flows. Digital technologies allow trash to be used as a resource and maintain data together with substances throughout the cycle through the notion of the "digital twin" [65]. Real-time data on the location, state, and availability of an item may be obtained via digital solutions, which can also improve material traceability, facilitate access to goods and services, and streamline procedures [66]. Additionally, data analytics may be used as a tool to purchase replacement parts, plan maintenance, minimize production downtime, forecast product condition and wear, and optimize energy use [67]. Digital and online platforms and technology help CE by fostering stronger relationships among supply-chain participants. An ecosystem, rather than a single company, often closes the loop in a model of circular business [62]. Enhancing resource efficiency and improving accessibility to data and utilization within the supply chain may be achieved through both horizontal and vertical integration of stakeholders throughout the product lifetime, beginning with the producer of raw materials and concluding with the end user [68].

11.3 CORPORATE INITIATIVES AND TRANSFORMATIONS

The digital revolution and the circular economy are happening simultaneously, and cooperation between the two is necessary for a successful shift that will solve global issues. According to the European Commission, this is a "twin transition" in which the digital and environmental transitions must support one another [69]. However, it is commonly understood that digitalization might have unforeseen implications, such the need for large quantities of energy [70]. Thus, it is critical to comprehend how digital technology may support a flourishing circular economy, for instance, by facilitating resource exchanges, teamwork, and online platforms for recycling and reuse [71]:

- In what ways may companies use digital technology strategically to expedite their internal shift to a circular economy?
- How might digital platforms and technology foster ecosystems for the circular economy and collaboration?
- How might "circular economy leapfrogging," particularly in the Global South, be made possible by digital technology?
- What are the drawbacks of digital technology in terms of circularity, and how may such drawbacks be addressed?

We welcome theoretical contributions as well as actual instances using action-based methodologies. Strong theoretical underpinnings accompanied by interdisciplinary methods that integrate with engineering, design, policy, and the environmental sciences are also highly welcomed.

11.3.1 IKEA's Circular Value Chain

With a goal of being completely circular by 2030, international furniture retailer IKEA has been leading the way in incorporating circular economy ideas into its value chain. The company's strategy includes encouraging product take-back and program recycling, employing recycled and renewable resources, and producing long-lasting goods. IKEA has extended product lifespans and decreased waste by introducing modular furniture that is readily updated, repaired, and personalized [72]. Moreover, they should promote a cyclic existence of their products, IKEA's "Second Life for Furniture" program allows the buyers to take back the used furniture which, in turn, is renovated and resold [73]. To minimize its external cost, the corporation also spends on innovative materials such as ecological wood and bioplastics [74]. In addition to promoting preservation, this has contributed toward major improvements of the overall environmental sustainability, thus boosting the company's brand image to its clients. IKEA is an example when large companies might be given a role model which fosters the changing of Oops! economy that is circular by including circularity as a key component of their primary business model [75].

11.3.2 Take-Back and Refurbishment Programs

Take-back of goods is part of the CE which is promoted as a sustainable product to the old linear economy which is still present at the helm of the new consumption model. This chapter thus looks at how CE has increased the focus on the regeneration and restorative aspects can be juxtaposed to the take–make–use–dispose model inherent in the linear economy. According to estimates, CE may generate about 700,000 new employment and boost the GDP of the European Union by 0.5% in a European setting [76]. Take-back programs are said to be crucial in helping the CE close the loop between production and end-of-life in order to realize the use of residual product value [24]. The industry faces difficulties in creating commercially viable take-back systems, despite the potential advantages of CE and growing scholarly interest in this field [77]. For example, Hvass and Pedersen (2019) advocate for research into the creation of commercially viable take-back system-based business models [78]. One of the things preventing CE procedures from being widely adopted is this. This study's goal is to identify the variables affecting the product take-back systems' financial success in order to address and investigate this possibility.

Consequently, the technical material and manufacturing processes of a CE revolve around the practices that involve direct reuse, repair, refurbishing, and remanufacturing (together referred to as "Value-Retention Processes" or "VRPs") [79]. We use the VRP criteria provided by the UN Environment program's International Resource Panel (IRP) for the purposes of this research (2018): Straightforward Reuse—the utilization of a non-waste product once again for its intended use without the need for maintenance or restoration; Repairing—a product involves either replacing its broken components or repairing a specific flaw to restore it to working order so that it may be utilized for its intended purpose; Refurbishment—the process of altering a product during servicing or intermediate maintenance procedures in order to improve or restore functionality and/or performance or to comply with relevant technical standards, resulting in the creation of a fully functional product that can be used for at least the original intended purpose; and Remanufacturing—this is a standardized industrial procedure used in factories or other industrial settings to return cores to their prerestoration state, if not better [80].

11.3.3 Designing for Recycling and Repurposing

The fact that the present aims are focused on a few areas—recycling, recovery, and resource efficiency—is emphasized in Figure 11.2. However, a CE includes more potential solutions than efficiency (e.g., product-service systems, or PSS) and a broader variety of techniques than recovery and recycling (e.g., reusing or refurbishing) [81, 82]. The fact that the current objectives are so limited suggests that new goals are desperately needed in order to more accurately capture the complicated nature of a CE. This implies that the examination of both new and current targets should take a distinct approach. Examining the scholarly and gray literature on targets and CE can help you evaluate the current targets. Alternatively, new goals may be split into two categories: newly created ones, like the ones I suggest in this study, and those that have already been suggested but not yet realized (i.e., promoted by academics and experts). Reaching these goals is necessary to fortify a CE completely. New objectives may be created by evaluating the requirements, advantages, issues, and potential solutions of any business aiming to be circular. Suggested targets can be analyzed, but they can also be enlarged or reworked as needed [83].

11.4 AUTOMOTIVE INDUSTRY

According to UN DESA (Population Division), the number of people on Earth is predicted to reach about 10 billion by 2060. The middle class's proportion of the world population is predicted to grow at the same time (HKExnews), which will probably lead to a sharp growth in demand for goods in the automobile industry, particularly in emerging nations. In addition, the automobile industry's shift from traditional to

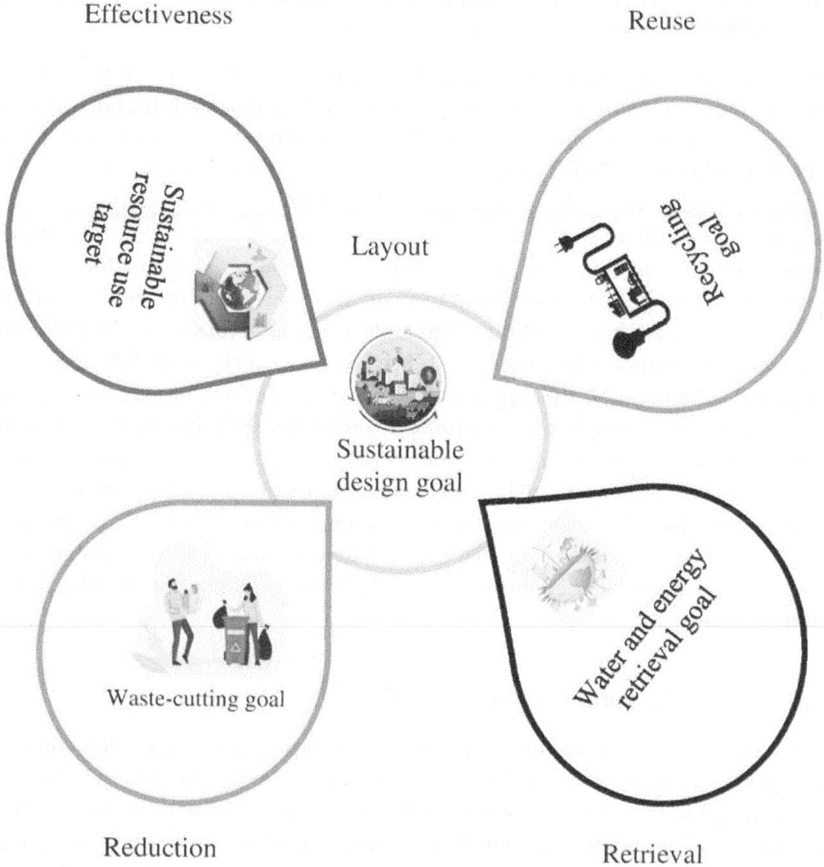

FIGURE 11.2 Primary current CE objectives broken down by application areas.

novel drive train technology is a contributing factor to the rising need for nonrenewable resources, such as so-called essential minerals (IEA) [84].

By using fewer resources, CE reduces waste, pollution, and energy required for manufacturing in the automobile sector. Increased growth and economic success of society as a whole are associated with the use of more ecologically friendly products that lessen their detrimental effects on the natural world and CO_2 emissions [85]. Thus, the primary driver for the implementation and accomplishment of the sustainable development goals on a worldwide scale is the CE's vision to raise awareness and desire to act from both government authorities and industrial enterprises, including the automobile sector [86]. With businesses becoming more conscious of environmental issues, energy conservation, and mitigating climate change, CE practices may also benefit manufacturing companies' sustainable supply chain management. A manufacturing firm's supply chain partners must be involved in helping it achieve its environmental conservation and pollution reduction objectives, in addition to the company itself [87].

11.4.1 Renault's Remanufacturing Operations

Thanks to its vast remanufacturing facilities, Renault has been a forerunner in the automobile industry's shift to a circular economy. The firm specializes in the remanufacturing of motors, transmissions, and other automotive parts at its Circular Economy Plant in Choisy-le-Roi, France. This facility drastically lowers waste and the need for raw resources by disassembling and refurbishing used components while making sure they adhere to the same requirements as new ones [29]. To strengthen the organization's longevity and performance, new materials and technologies are often incorporated in the time-consuming inspection, cleaning, and reconstructions of components that constitute Renault's remanufacturing chain [88]. Since remanufactured components consume 88% less raw material and 80% less energy as compared to new parts, this approach is not only effective in extending product life cycles but it also reduces environmental pressures [89]. In addition, Renault's circular business techniques have proven to have financial performance grandeur: Renault's remanufacturing branch, for instance, not only provides consumers with cheaper options but also accounts for vast revenue returns [90]. These programs demonstrate how remanufacturing can play a role in the use of resources and can act as a key element in circular economy strategies in the organization, particularly the automobile industry.

11.4.2 Economic and Environmental Benefits

The proposed approach based on the practice of a circular economy offers numerous opportunities for the economy and the environment. From the business point of view, practices linked to circular economy can allow reaching considerable savings for an organization as it decreases the demand for raw materials and expenses connected with disposal of waste. For instance, firms can repurpose products to gain extra value by recycling, remanufacturing, or material reuse, this not only extends product life but also hedges on volatility of material costs. According to figures from the Ellen MacArthur Foundation, a transition to circular economy may generate a $1 trillion net saving by 2025 [91]. As for environmental contributions, circular economy means less waste and resource consumption, fewer emission of greenhouse gases, and delay of environmental depletion. Circular processes help in the measures which aim to decrease emissions, conserve the resources, and minimize the influence of consumer and production activities by breaking the reliance on the product life cycle. For example, through remanufacturing of old auto components into new ones, the facility of Renault Group remanufacturing has significantly reduced the adverse impact of the company's products on the natural resources preserved and the energy utilized in developing new products [29]. Second, the diffusion of CE concepts at a large scale will lead to the reduction of CO_2 emissions by up to 50%, thereby supporting sustainable development as well as assisting in achieving the global climate goals. The global need for enterprises to adopt economic circles is justified by the advantages that go hand in hand with economic and environmental returns. Those organizations continue to be more adaptable and successful and at the same time contribute to an emergence of an economy that is sustainable and also friendly to the environment.

11.4.3 Case Study: Choisy-le-Roi Plant

The company's plant at Choisy-le-Roi in France is an excellent example of how circular economy concepts might be implemented in the automotive field successfully. Creating new engines, gearboxes, and injector pumps for automobile industries and reconditioning of automotive engines, turbo chargers, and other related auto parts is the specialties of this facility. Under its operating mode of remanufacturing with the Choisy-le-Roi facility, the life cycle of automobile parts is prolonged while the environmental impact of manufacturing new component is significantly reduced. Another improvement estimated based on the data obtained is the cost-effectiveness, which is one of the main economic benefits observed in the investigated area of Choisy-le-Roi. Comparing to fabrication of new component, this remanufacturing plant consumes only 12% of the raw material and energy. Apart from resulting in huge cost-savings for Renault, such reduction translates to improved sustainability of the supply chain [28]. In addition, the operation helps to cut garbage production and the need for landfill area by gathering and recycling resources.

As for the environment, it is enough to concentrate on the circle processes that have a positive impact—for example, the Choisy-le-Roi factory. As to the impacts of the plant activities on the overall consumption, production, and emission reductions, the plant's activities show 70% less water consumed than the conventional production methods and cutting chemical output by 60%. In addition, due to this aspect of the plant's practice to reuse components, the organization has cut down on CO_2 emission, thus responding to the global struggle against climate change [29]. Not only does Renault is friendly toward the environment, it also enhances its image and beats the market performance by integrating refurbishment in business strategic unity.

Fundamentally, circular economy manufacturing strategies must not be analyzed only in the light of their performant implementation context at the Choisy-le-Roi plant, as evidence proves that they generate actual competitive advantages. It also shows how other sectors may indeed implement the fiduciary duty to make money and protect the environment at the same time.

11.5 ELECTRONICS AND CONSUMER GOODS

Consumer electronics have revolutionized home office operations, communication, and entertainment experiences in the last several decades. Examples of these include tablets, wireless headphones, and smart phones. [92, 93]. The quick access and sharing of data and information has been made possible more recently and in a very short amount of time by the growth of cloud computing and enhanced device connectivity through Internet technologies. Additionally, new entertainment mediums like streaming and Internet-based gaming have gained popularity [93, 94].

The consumer electronics industry category, which includes a wide range of other items, including televisions, pills, disposable digital cameras, headsets, audio devices, and more, frequently includes mobile phones (see, for example, Refs. [92, 93]). In Europe, the total amount spent by households on consumer electronics was 1.32% in 2017. Of that amount, the manufacturing industry generated approximately €60 billion in revenue. The manufacturing of computer, electrical, and optical devices employed little more than 1.1 million individuals in the EU in 2015—the lowest number of workers in the industry in recent memory. More specifically, the production of communication equipment has all but vanished from Europe; in Germany, for instance, employment in this sector has decreased by 30% [95].

Many people agree that the circular economy has the capacity to address the social and environmental issues of today that are brought on by years of excessive economic expansion. The circular economy is an economic model that distinguishes between technological and biological cycles and aims to maintain goods, components, and materials at their maximum utility and worth at all times. It is rejuvenating and regenerating by design. In particular, the CE is receiving more and more attention in the consumer products industry, which is an important development given the massive amounts of waste produced worldwide and the high consumption of agricultural output. However, the CE still continues being a rather fresh concept that will require further development. Namely, consumer products industry requires reliable CE indicators that provide insights on CE effectiveness and practical tips on further improvement among industry participants, as well as acceleration of circularity transition in the industry [96].

11.5.1 HP's Closed-Loop Recycling

HP Inc. has exhibited good governance in the integration of the concept of circular economy at the company by coming up with closed-loop recycling. This way, there is a cut on the current impacts on the environment; the gadgets are made from reused components from other gadgets that have been discarded. Its closed-loop recycling involves collection of hardware and cartridge returned by customers, and separation of necessary constituents to create new hardware and cartridges. It dramatically reduces occasions where raw resources are required while at the same time minimizing on the creation of trash. It means that HP's closed-loop waste disposal program has been proved as the financial sensible action. This way HP can reduce cost of expenses on throwing away trash as well as on obtaining raw resources through recycling commodities. The business also benefits from improving its brand image as the innovator in sustainability and developing the client base. This strategy means creating new opportunities for commercial growth in the growing environmental products market as well as helped HP to minimize the value hazard linked to resource scarcity [29]. From an environmental perspective, HP's closed-loop recycling program has led to a drastic decrease in energy consumption and emission of greenhouse gases in the production of the products. For instance, the use of reused plastic in HP goods has cut down a large carbon footprint within the firm. Moreover, it has reduced pollution within the environment by having prevented the discharge of millions of cartridges into the dumping sites. The main interest in sustainability and product life cycle management illustrates how, using the circular economy techniques, HP demonstrates that companies can help the environment while improving their revenues [28]. Circular economy is one of the methodologies that can be incorporated into business strategies, in this case, one of the most successful cases of business application is the closed-loop recycling program of the company named HP. This case study underscores the need for creative recycling activities in constructing a stronger and sustainable economy.

11.5.2 Circular Practices in Electronics

Currently, electronics producers are adopting circular economy solutions to the scarcity of resources and the environment, despite working in a sector with rapid development and short useful life of their products. For circularity in this industry, the objectives focus on the optimization of product lifespan and promoting the recycling rates in order to decrease the generation of e-waste. In this case, businesses are implementing measures like designing for disassembly, using recycled material, as well as putting in place a take-back system.

Of the cases, Dell Technologies is perhaps one of the most celebrated firms that make use of circular economy ideas in their design and production functions. Due to easy disassembly of Dell's products, strategic material components can be recycled and reclaimed in the most efficient way. By employing recycled polymers from old electronics to form new parts, the recycling program of the company known as "closed-loop recycling" minimizes the need for virgin resources and reduce the impact that production has on the environment [97]. In addition, customers are urged to return their unused devices to Dell through its take-back program with a promise that the products are recycled or refurbished as necessary. Another emerging sustainability approach is circular economy, to which Apple duly committed. Apple has developed innovative recycling technology, including the Daisy robot, which takes apart iPhones with the purpose of recycling parts that contain valuable metals like rare earth elements, gold, and cobalt. This reduces the extent of impact of mining and processing of substance as well as the conservation of resources. These programs have made Apple's contribution up to nearly 48,000 metric tons of electronics in the same year since 2015 [98]. Hence circularity of processes in the electronics industry accrues financial benefits as well as environmental gains. From the organizational economic perspective, companies may reduce costs associated with wasted products and resources used to source raw materials. This way you may reduce the environmental cost of electrical items, conserve natural resources, and mitigate the effects of greenhouse gas emissions. As an industry capable of generating significant positive impacts on both the environment and economy, electronics sector has to adopt circular economy concepts as a structured approach to increase business resilience and sustainability [99].

11.5.3 Case Study: Printer Cartridge Recycling Program

Embracing Global Environmental Management Initiative—Printer Cartridge Recycling Program is another example of circular economy initiatives taken in electronics industry, that is, HP Inc.'s printer cartridge recovery. With the help of this application, discarded cartridges may be recovered. After being dismantled and cleaned, the cartridges can be recycled or refilled. Because of the program's sustainable strategy, HP is able to save a considerable amount of money and increase customer loyalty by reducing its need on virgin raw materials. The scheme has effectively decreased energy use and greenhouse gas emissions, saving a billion cartridges from winding

TABLE 11.1
Various Programs of Printer Cartridge Recycling

Aspect	Details	Reference
Economic Benefits		
Cost-saving measures	Production costs can be reduced by reducing dependence on virgin raw materials	[100]
Customer loyalty	Brands that put sustainability first attract environmentally conscious consumers	[100]
Closed-loop purchasing	Effective component recycling from cartridge returns	[100]
Benefits to the Environment		
Cutting down on waste	Stopped filling landfills with millions of cartridges for disposal	[99]
Energy conservation	Less energy is utilized when consumables are recycled as opposed to being produced brand-new	[99]
Diminished greenhouse gas emissions	Significant carbon footprint reduction for the company through recycling	[99]
Conservation of resources	Recycling plastic to create new cartridges protects natural resources	[99]
Effects of the Program		
Total cartridges recycled	More than 875 million cartridges are being recycled since the program's inception	[99]

up in landfills. With over 875 million cartridges recycled since its founding, HP has had a major beneficial economic and environmental effect (Table 11.1) [99, 100].

11.6 MANUFACTURING AND PRODUCTION

While they contribute to ecological issues, manufacturing companies—defined as businesses that employ machinery to make items from beginning materials and/or components—play a significant role in raising people's standards of living. The manufacturing sector is at the center of these issues because of the serious environmental consequences associated with unsustainable patterns of production and consumption. Since the 1987 Brundtland Commission established the idea of sustainable development as a method of development that can meet the needs of the world's population today without endangering the planet's ability to fulfill the needs of future generations, sustainability has gained significant traction [101]. The Brundtland Commission's recommendations remain applicable. With its five main pillars of sustainable development—people, planet, peace, prosperity, and partnership—the United Nations' 2030 Agenda for Sustainable Development lays out a course of action with 17 Sustainable Development Goals (SDGs), 169 targets, and 232 indicators to monitor progress [102]. The SDGs include a wide range of sustainable development topics, many of which are important to manufacturing organizations.

11.6.1 Integration of Circular Economy in Manufacturing

Manufacturing is changing as a result of the application of circular economy concepts, which place a strong emphasis on waste reduction, resource efficiency, and sustainability. This method includes creating goods that are long-lasting, recyclable, and reusable in addition to putting systems in place that allow resources to be reused indefinitely. Manufacturers may minimize their environmental impact and realize huge economic benefits by implementing circular processes.

Designing with disassembly and modularity in mind is a crucial step in implementing the circular economy in manufacturing. Products are made to have readily replaceable or upgradeable individual parts, which lowers waste and increases product longevity. In an effort to popularize sustainability and reduce costs, firms like Philips have developed electronics that can be upgraded or refurbished instead of discarded as a whole unit [103].

The adaptation of fresh and recycled products in production is another strategy of environmental management. By so doing, some natural resources are conserved and the energy and carbon emissions arising from extraction and processing of raw materials are saved. For example, Interface, a global manufacturer of commercial flooring, was initially a prime example of a company which incorporated recycled parts into its materials eradicating almost all negative influence on the environment and setting a precedent for leeward manufacturing processes [104]. In addition, it is understood that manufacturers are using reverse logistics and take-back programs to a larger extent for the recovery of end-of-life products for disposal and remanufacturing. This closed-loop technology always feeds the resources back into manufacturing, potentially reducing waste and the need for fresh material. One particular example is the automobile industry in which established take-back systems let companies, for example, BMW, to refurbishment or recycle automobile parts in fact exemplifying the practical realization of CE ideas in a complex industrial sector [105].

Manufacturing that incorporates circular economy principles promotes innovation and competitiveness in addition to sustainability. Manufacturers may build robust business models that support both economic and environmental objectives by reconsidering product design, material selection, and end-of-life management. This will pave the way for an environment-friendly industrial future.

11.6.2 Smart Factories and Industry 4.0

Industry 4.0, also referred to as the technological revolution, includes big data, cybersecurity, cloud computing, Internet of Things (IoT), simulations, integration systems, additive manufacturing, big data, and augmented reality as the primary broad factors of technological innovations work aimed at constant enhancement. Industry 4.0 is a concept that integrates manufacturing and production processes with ICTs, or information and communication technologies [106]. The main technologies that are integrated into Industry 4.0 are summarized in Figure 11.3, which also lists the Industry.0 pillars.

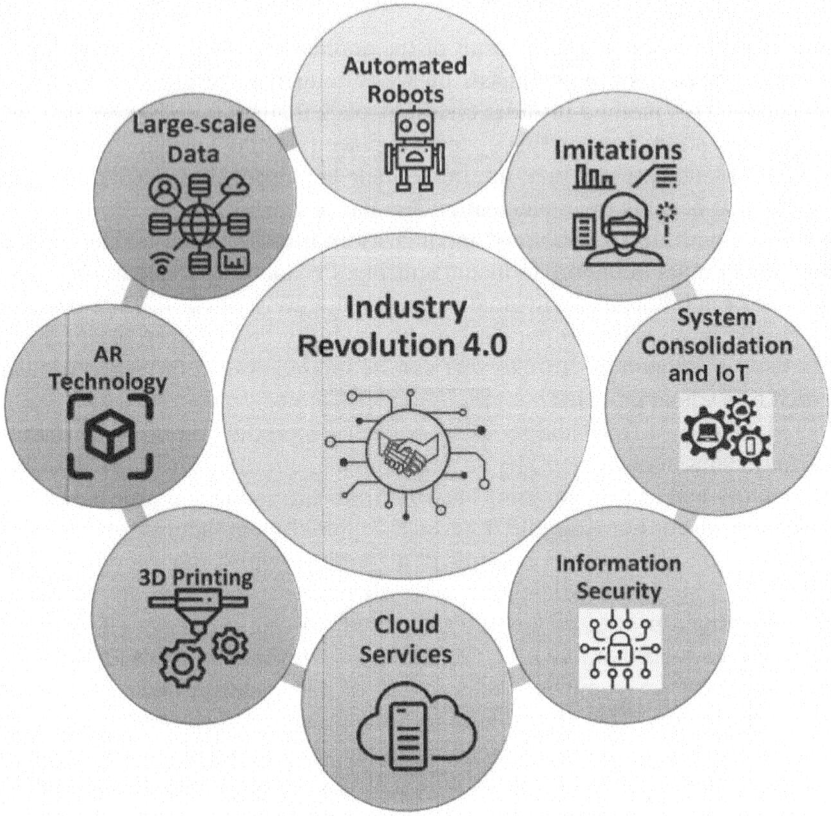

FIGURE 11.3 Industry 4.0 pillars.

11.6.3 Robotics in Quality Control and Maintenance

Modern industrial processes, including assurance of quality and maintenance, now heavily rely on robotics to increase productivity, precision, and dependability. Robots are used in quality control to carry out very accurate measurements and inspections that are beyond the capability of humans. Modern robotic systems with high-definition cameras and sensors are able to identify even the smallest product flaws, guaranteeing constant standards of quality. For example, robotic vision systems are used by automakers to check automobile bodywork for flaws, and they do it more quickly and accurately than with manual inspections [107].

Robots in maintenance are changing the way that equipment is maintained. Robotics and artificial intelligence (AI)-driven predictive maintenance use sensors along with information analytics to forecast equipment failure dates and schedule repair before problems develop. By doing this, downtime is reduced and machinery longevity is increased. In dangerous situations, robots can also carry out maintenance duties, lowering the risk to employees. Robots, for instance, are employed in the oil and gas sector to examine and maintain pipelines, sometimes in hazardous or challenging-to-access areas for people [108].

Robotics integration in maintenance and quality control not only increases operational effectiveness but also dramatically lowers expenses related to faulty goods and downtime. Furthermore, the steady performance and fatigue-free operation of robots contribute to increased productivity. It is anticipated that as technology develops, robot's integration in these fields will become even more complex, providing more precision and predictive power [109].

11.7 MATERIALS SCIENCE

The broad area of materials science examines the characteristics, functionality, and uses of materials. It combines ideas from engineering, chemistry, and physics to comprehend and control the molecular and atomic makeup and characteristics of materials. New materials with improved qualities and functionalities have been developed as a result of the substantial influence that materials science breakthroughs have had on a number of sectors.

The simplest layer of carbon atoms organized in a two-dimensional nature honeycomb lattice is called graphene, and its creation is one of the most significant advances in materials research. Due to its remarkable mechanical, electrical, and thermal characteristics, graphene is a material that has great promise for use in composite materials, electronics, and energy storage, among other fields [110]. Scientists are always looking for new methods to use the special qualities of graphene to transform business and technology.

Environment-friendly and sustainable material development also forms another aspect of materials science. For instance, the innovations made in biodegradable composites and polymers may reduce the quantities of plastics generated, which in turn reduce the ill effects on the environment. In due course, these materials can decompose naturally and help to minimize pollution and encourage the emergence of an eco-friendly society [111].

Materials science has made major contributions to the medical industry by advancing the area of biomaterials, which are used to create prostheses, implants, and medication delivery systems. Biomaterials innovations have enhanced medical device functioning and compatibility, improving patient results and quality of lives. One notable achievement in cardiovascular therapies is the creation of bioresorbable stents which disintegrate after serving their intended purposes [112].

All things considered, the field of materials science keeps encouraging innovation and advancement in a variety of fields, supporting the creation of novel technologies and answers to global problems.

11.7.1 AUTOMATED MATERIAL SYNTHESIS AND TESTING

The circular economy has undergone a revolutionary change with the introduction of automated substance synthesis and testing, which has improved the accuracy and efficiency of material creation and evaluation procedures. The creation of novel materials and their subsequent evaluation under many conditions are accelerated by these automated systems, which make use of sophisticated robotics, AI, and machine learning algorithms. Through efficient utilization of the beginning components and a reduction in synthesis mistakes, this combination not only expedites the search for sustainable resources but also minimizes waste. High-throughput experimental platforms, for example, have the capacity to test dozens of material combinations concurrently, which greatly reduces the amount of time needed for conventional trial-and-error techniques [91].

Furthermore, AI-powered predictive modeling can anticipate the characteristics and functionality of novel materials, directing the synthesis process in the direction of the most likely candidates [113]. The development of materials that are biodegradable and recyclable is greatly aided by these developments, which will eventually lead to a more circular and sustainable economy [114].

11.7.2 Nanotechnology and Advanced Materials

Advanced materials and nanotechnology have the power to revolutionize a number of sectors and usher in a new period of creativity and economic expansion. Faster and more potent gadgets are the result of the ongoing shrinking of electronic components made possible by nanotechnology [115, 116]. Cutting-edge materials with exceptional mechanical strength and electrical conductivity, such as graphene and carbon nanotubes, are excellent choices for the next-generation electronics [117]. In the health care industry, nanotechnology has completely changed imaging, medication delivery, and diagnostics. Treatments can be more precisely and successfully administered by using nanoparticles that are engineered to target particular cells. Prosthetics, tissue engineering scaffolds, and biocompatible implants all have been made possible by advanced materials [118]. Batteries, energy storage devices, and solar cells all have become more efficient because of nanotechnology. Modern materials, such as perovskite solar cells, provide an affordable substitute for conventional photovoltaic systems. In the energy industry, nanomaterials also improve the efficiency of energy storage systems. Advanced materials with remarkable mechanical qualities are used in manufacturing to help create lighter and more durable materials for the automotive and aerospace sectors. Coatings based on nanotechnology improve corrosion resistance and wear resistance, extending the life of devices and machinery. Environmental issues are addressed by both nanotechnology and novel materials [119, 120].

11.8 APPLICATIONS OF FUZZY LOGIC IN CIRCULAR ECONOMY

In the topic of circular economy, fuzzy logic—a type of many-valued logic that deals with approximations instead of fixed and accurate reasoning—has found extensive use. This method works especially well for addressing the complexity and inherent unpredictability of circular economic systems. Fuzzy logic, for example, may be used to assess many factors, including waste kind, recycling potential, and impact on the environment, in order to optimize processes for handling waste [121]. Decision-makers may evaluate and rank waste treatment choices more effectively by using fuzzy logic algorithms, which improves recycling and recovery of resources procedures [122].

Also, because of the uncertainty involved in predicting demand forecast, supplier reliability, and production capacity, the developed fuzzy logic approach that underpins the sustainability of supply chain design can enhance the relevant supply chain functionality [123]. All these approaches promote continued flow of material through the supply chain, hence increasing the robustness and flexibility of supply chain techniques. Overall, fuzzy logic makes decisions stronger which in turn, assist with sustainable development goals when implemented into circular economy projects.

11.8.1 Overview of Fuzzy Logic

There was Lotfi Zadeh who introduced fuzzy logic in 1965 as formulate means to cope with the concept of partial truth, that is the truth value can be ranging from fully true to fully false. Fuzzy logic variables are those which contain truth values between 0 and 1; whereas the general variables make use of 0 and 1 only [124]. The use of fuzzy logic is advisable when it is vital to solve complex real-life problems because of their inherent vagueness due to the fact that it does not require impressive accuracy. In true fuzzy logic systems, fuzzy sets and rules that are similar to human thinking are employed in order to mimic uncertainty and imprecision [125]. Due to its ability to proceed with the uncertainty and provide the right results in the issues that cannot be solved by the classical logical way, this method is widely used in different areas, such as the management systems, the pattern recognition, and decision-making [126]. Fuzzy logic is a perfect match for the circular economy because with its help, the industry can use more superior approaches to waste minimization, resource utilization, and environmentally sound practices.

11.8.2 Fuzzy Logic Models in Waste Management

Consequently, for raising the effectiveness of waste management systems in the context of circular economy, fuzzy logic models are essential. Due to these uncertainties and complexities involved in waste formation, categorization, and processing, these models employ fuzzy set theory. In decision-making, policymakers come up with what could be more precise and more efficient waste management plans that are more likely to consider fluctuations in waste composition and generation frequencies by the use of fuzzy logic. For instance, the applying of a fuzzy logic–based technique can help sort and recycle wastes of different types of product, contaminant concentration, as well as recovering potential based on the characteristics of waste streams. Improved resource recovery and limited landfill use is the outcome of this and this aligns with the circular economy important tenets of waste elimination and sustainable use of resources. Studies have shown that using the approaches based on fuzzy logical models yields better and more accurate outcomes in managing waste and helps solve transition to circular economy [127].

11.8.3 Optimizing Resource Recovery Pathways

For these reasons, enhancing the circular economy's resource recovery paths must be facilitated. Cohesion and economic activities within the business and communities involved could immensely enhance the sustainability of the activities by improving on the efficiency of these corridors. Optimization is achieved in one way by regulating the process that pass elements from the inputs to the outputs through the recovery processes enhanced by modern technology such as machine learning and IoT. To facilitate sorting and recycling processes in order, for example, real-time information about the state of the material or the conditions for processing may be received using specific sensors and algorithms for processing received data. Moreover, aware of the impacts of different recycling and reuse options, life cycle assessment (LCA) technologies can help decide on the recovery paths that are most beneficial from the environmental point of view. The integration of the best practices of the sorting technology with the LCA outcomes can provide a decrease in the estimate of the environmental effects by 20% and in the matter recovery rates by 20% based on the case study to improve the electronic debris recovery routes [128]. Such changes provide an insight of importance of adopting technology and improving it in order to attain the goals of circular economy.

11.9 SECTOR-SPECIFIC CASE STUDIES

In order to create a circular economy that can enhance connected business ecosystems, the economy of circularity also centrally comprises several sectors to do with biotechnology and medicine, energy and utility, and food systems and agriculture. Implementations of circularity are present in the food systems and farming by using waste agricultural products to manufacture helpful products like biopesticides and biofertilizers. This enhances the returns on the crops and fertility of the soil besides reducing wastage to enhance food security [129]. The generation of biogas from organic matter and agricultural leftovers offers a sustainable energy source to the power and utilities industry, reducing dependency on petroleum and coal and emissions of greenhouse gases [130].

Initiatives to transform food waste from cities into biogas, it is then utilized to power nearby villages or returned to agriculture as nutrient-rich digestate, are an example of this synergy [131].

Circular concepts are also being adopted by the biomedical and pharmaceutical sectors, which are creating procedures to extract and recycle valuable components from biological waste. Pharmaceutical by-products, for instance, can be used as fuel for the generation of bioenergy or in agricultural applications [132]. Furthermore, biotechnology breakthroughs make it possible to produce bio-plastics and bio-based products from agricultural waste, which may be applied to a variety of sectors and lessen reliance on petroleum-based goods [133].

There are substantial economic and environmental benefits to the melding of these industries in the Netherlands, according to a case study. The project produced a 25% increase in the generation of renewable energy and a 30% decrease in waste by deploying a closed-loop system that turned agricultural waste into biogas and bioproducts. It also encouraged innovation in applications of biotechnology [134]. This is a prime example of how a circular economy may build interconnected systems that support sustainability in a range of sectors.

11.10 CHALLENGES AND OPPORTUNITIES

Making the shift to a more circular economy comes with a number of possibilities and big problems. The necessity for systemic transformation across sectors and industries, which necessitates large investments in new infrastructure, business models, and technology, is one of the main obstacles [4]. In order to promote the broad adoption of circular practices, it is also necessary to remove regulatory and policy hurdles. For example, the policies governing waste management already in place could not encourage creative recycling or reusing techniques, which would impede advancement [27].

The intricacy of creating circular processes and products is another difficulty. This frequently entails reevaluating conventional design concepts to guarantee that goods are readily recyclable, dismantled, and mended. Since the upfront costs of these changes may be significant and the return on investment from such changes may not be obvious, immediate, or certain, the economic consequences of such changes must be effectively and efficiently managed [135].

11.10.1 Future Prospects and Emerging Trends

Governmental measures, as well as customers' and industries' attitudes, shift toward the circular economy, and the technologies' advancements guarantee the promising future and trends for the circular economy. Use of advanced technologies such as blockchain, artificial intelligence, and the Internet of Things in enhancing the circularity, efficiency and acceptance of the circularity concept is one of the major trends. These technologies enable generation of new innovative services such as product-as-a-service (PaaS), better supply chain, and better ways of managing resources [24]. Another is the shift to circular design, meaning the primary design of objects with lifespan, modularity, and de-constructability from scratch, for recycling. These guidelines are slowly finding their way into businesses more and more since this way, it will be easy to disassemble the product at the end of its life cycle. Development of longer lasting materials for products and designs for recycling are emerging as greater improvements come in materials science [136]. Also rising in prominence as one of the key pillars of the circular economy is the bioeconomy. To produce sustainable materials, energy, and products, biological stocks and processes are employed. The latest developments in biotechnology have enabled the formation of biodegradable plastics, biofuels, and many other biomaterials that are more environment-friendly than materials produced from petroleum-based products. Thus, the successful application of the bioeconomy principles should be targeted at such sectors as agriculture, food industry, and pharmacology [137].

Frameworks for regulations and policy are transitioning to also accommodate for a transition in the circular economy. In order to promote circularity among companies, governments and international groups are, for instance, developing eco-design standards and a circular procurement policy as well as implementing EPR. These regulations provide a market nice for players that are committed to circularity while at the same time increasing sustainability [138].

Also, based on the growing consumers' awareness, there is a beaconing trend for more circular products and solutions. As people become more conscious of how linear consumption habits affect the environment, they become more interested in recycling, sustainable products, and waste reduction programs. Companies are being compelled by customer demand to develop and implement circular strategies in order to fulfill market expectations and stay competitive [139].

All things considered, the circular economy has a bright future since players are always innovating and working together to create more robust and sustainable economic systems.

11.11 CONCLUSION

Achieving sustainability and financial resilience requires adopting a practical and essential strategy that goes beyond the theory of the circular economy. It is clear from the analysis of case studies and real-world applications that circular economy strategies may have a major positive impact on the environment, the economy, and society. They include wastage minimization, efficient resource management, and availability of new business opportunities. This change is accelerated by the combination of cutting-edge technology like blockchain, IoT, and AI with creative business models and encouraging regulatory frameworks. For widespread acceptance to occur, however, obstacles pertaining to systemic shifts, regulatory hurdles, and early expenses must be addressed.

The condition indicates which elements should be followed in everyday practice since several applications in real life prove the effectiveness and possibilities of circular concepts in various fields, such as biotechnology, energy, and agriculture. For instance, transforming agricultural waste such as crop residues into bioenergy and biogoods creates value out of waste while at the same time creating high value-added commodities and renewable energy. On the same note, applying circular design principles in production promotes the endurance, modularity and reusability, and recyclability of the products that are produced. More research needs to be done to unlock the full potential of circular economy and overcome present challenges in view of continuously growing consumers' awareness and interest toward sustainable products, on the one hand, and constant improvements in firms' and governments' cooperation on the other. Future prospects are bright because of the continuous progress in consumer behavior, policy, and technology that is propelling the transition to a more robust and sustainable economic system. Adopting the circular economy is crucial to tackling today's urgent environmental issues, promoting social progress, and economic expansion.

REFERENCES

1. Ready-Mulvey, G., and W. R. Stahel. *The Potential for Substituting Manpower for Energy: Final Report for the Commission of the European Communities*. Geneva: Battelle, Geneva Research Centre, 1977.
2. Ellen MacArthur Foundation. *Towards the Circular Economy: Economic and Business Rationale for an Accelerated Transition*. Cowes, Great Britain, 2013. https://www.ellenmacarthurfoundation.org.
3. Murray, Alan, Keith Skene, and Kathryn Haynes. "The circular economy: an interdisciplinary exploration of the concept and application in a global context." *Journal of Business Ethics* 140, no. 3 (2017): 369–380. https://doi.org/10.1007/s10551-015-2693-2.

4. Kirchherr, Julian, Denise Reike, and Marko Hekkert. "Conceptualizing the Circular Economy: An Analysis of 114 Definitions." *Resources, Conservation and Recycling* 127 (2017): 221–232. https://doi.org/10.1016/j.resconrec.2017.09.005.
5. European Commission. *Towards a Circular Economy: A Zero Waste Programme for Europe*. Brussels: Communication from the Commission to the European Parliament, the Council, the European Economic and Social Committee, and the Committee of the Regions, 2014. https://www.municipalwasteeurope.eu/sites/default/files/COR-2014-EN.pdf.
6. De Romph, T. J. *The Legal Transition Towards a Circular Economy.* Dissertation, Leuven: Faculties of Law of KU Leuven and Hasselt University, 2018. https://www.law.kuleuven.be.proxy.library.uu.nl/home/algemeen/agenda-2017-2018/doctoraatsverdediging-thomas-j-de-romph.
7. European Commission. *Single-use Plastics: New EU Rules to Reduce Marine Litter.* Brussels, 2018. https://ec.europa.eu/commission/presscorner/detail/en/IP_18_3927.
8. McDowall, Will, Yong Geng, Beijia Huang, Eva Barteková, Raimund Bleischwitz, Serdar Türkeli, René Kemp, and Teresa Doménech. "Circular economy policies in China and Europe." *Journal of Industrial Ecology* 21, no. 3 (2017): 651–661. https://doi.org/10.1111/jiec.12597.
9. Kirchherr, Julian, Laura Piscicelli, Ruben Bour, Erica Kostense-Smit, Jennifer Muller, Anne Huibrechtse-Truijens, and Marko Hekkert. "Barriers to the Circular Economy: Evidence from the European Union (EU)." *Ecological Economics* 150 (2018): 264–272. https://doi.org/10.1016/j.ecolecon.2018.04.028.
10. Dobbs, Richard, Jeremy Oppenheim, Fraser Thompson, Marcel Brinkman, and Marc Zornes. *Resource Revolution: Meeting the World's Energy, Materials, Food, and Water Needs*. McKinsey & Company, 2011. https://www.mckinsey.com/business-functions/sustainability/our-insights/resource-revolution.
11. Ellen MacArthur Foundation. *Towards the Circular Economy: Economic and Business Rationale for an Accelerated Transition.* Cowes, UK: Ellen MacArthur Foundation, 2013. https://www.ellenmacarthurfoundation.org.
12. Mathews, J. A., and H. Tan. "Progress Toward a Circular Economy in China." *Journal of Industrial Ecology* 15 (2011): 435–475. https://doi.org/10.1111/j.1530-9290.2011.00332.x.
13. Lehmann, M., B. De Leeuw, E. Fehr, and A. Wong. *Circular Economy: Improving the Management of Natural Resources*. Bern, Switzerland: World Resource Forum, 2014.
14. Xue, B., X. P. Chen, Y. Geng, X. J. Guo, C. P. Lu, Z. L. Zhang, and C. Y. Lu. "Survey of Officials' Awareness on Circular Economy Development in China: Based on Municipal and County Level." *Resources, Conservation and Recycling* 54 (2010): 1296–1302. https://doi.org/10.1016/j.resconrec.2010.05.010.
15. Su, B., A. Heshmati, Y. Geng, and X. Yu. "A Review of the Circular Economy in China: Moving from Rhetoric to Implementation." *Journal of Cleaner Production* 42 (2013): 215–227. https://doi.org/10.1016/j.jclepro.2012.11.020.
16. Singh, J., and I. Ordonez. "Resource Recovery from Post-Consumer Waste: Important Lessons for the Upcoming Circular Economy." *Journal of Cleaner Production* 134 (2016): 342–353. https://doi.org/10.1016/j.jclepro.2015.12.020.
17. Ngan, S. L., et al. Prioritization of Sustainability Indicators for Promoting the Circular Economy: The Case of Developing Countries." *Renewable and Sustainable Energy Reviews* 111 (2019): 314–331. https://doi.org/10.1016/j.rser.2019.04.020.
18. Ogunmakinde, O. E. "A Review of Circular Economy Development Models in China, Germany, and Japan." *Recycling* 4, no. 27 (2019). https://doi.org/10.3390/recycling4020027.
19. Ellen MacArthur Foundation. *Towards a Circular Economy: Economic and Business Rationale for an Accelerated Transition.* Cowes, UK: Ellen MacArthur Foundation, 2012.
20. Lacy, Peter, Justin Keeble, R. J. T. M. A. C. K. P. A. T. McNamara, Jakob Rutqvist, Tomas Haglund, Mengmeng Cui, Adam Cooper, C. Pettersson, E. Kevin, and P. Buddemeier.

"Circular Advantage: Innovative Business Models and Technologies to Create Value in a World without Limits to Growth." Accenture: Chicago, IL, USA 24 (2014). https://www.accenture.com/t20150523T053139__w__/us-en/_acnmedia/Accenture/Conversion-Assets/DotCom/Documents/Global/PDF/Strategy_6/Accenture-Circular-Advantage-Innovative-Business-Models-Technologies-Value-Growth.pdf. Accessed October 30, 2020.
21. Batista, L., M. Bourlakis, P. Smart, and R. Maull. "In Search of a Circular Supply Chain Archetype—a Content-Analysis-Based Literature Review." *Production Planning & Control* 29 (2018): 438–451. https://doi.org/10.1080/09537287.2017.1343502.
22. Bressanelli, G., M. Perona, and N. Saccani. "Challenges in Supply Chain Redesign for the Circular Economy: A Literature Review and a Multiple Case Study." *International Journal of Production Research* 57 (2019): 7395–7422. https://doi.org/10.1080/00207543.2018.1542176.
23. Lüdeke-Freund, F., S. Gold, and N. M. P. Bocken. "A Review and Typology of Circular Economy Business Model Patterns." *Journal of Industrial Ecology* 23 (2018): 36–61. https://doi.org/10.1111/jiec.12763.
24. Bocken, N. M. P., I. de Pauw, C. Bakker, and B. van der Grinten. "Product Design and Business Model Strategies for a Circular Economy." *Journal of Industrial and Production Engineering* 33, no. 5 (2016): 308–320. https://doi.org/10.1080/21681015.2016.1172124.
25. Gusmerotti, N. M., F. Testa, F. Corsini, G. Pretner, and F. Iraldo. "Drivers and Approaches to the Circular Economy in Manufacturing Firms." *Journal of Cleaner Production* 230 (2019): 314–327. https://doi.org/10.1016/j.jclepro.2019.05.044.
26. Kumar, V., I. Sezersan, J. A. Garza-Reyes, E. D. R. S. Gonzalez, and M. A. AL-Shboul. "Circular Economy in the Manufacturing Sector: Benefits, Opportunities and Barriers." *Management Decision* (2019). https://doi.org/10.1108/MD-09-2018-1070.
27. Lieder, M., and A. Rashid. "Towards Circular Economy Implementation: A Comprehensive Review in Context of Manufacturing Industry." *Journal of Cleaner Production* 115 (2016): 36–51. https://doi.org/10.1016/j.jclepro.2015.12.042.
28. Ellen MacArthur Foundation. "Make Fashion Circular." Accessed August 7, 2023. https://www.ellenmacarthurfoundation.org/our-work/activities/make-fashion-circular.
29. Renault Group. "Circular Economy Strategy." Accessed August 7, 2023. https://www.renaultgroup.com/en/our-commitments/circular-economy/.
30. Pires, A., G. Martinho, and S. Rodrigues. "Fuzzy Logic Model for Evaluating the Performance of Urban Solid Waste Management Systems." *Waste Management* 31, no. 6 (2011): 1355–1365. https://doi.org/10.1016/j.wasman.2011.01.023.
31. Chang, Ni-Bin, and Erika Davila. "Municipal Solid Waste Characterizations and Management Strategies for the Lower Rio Grande Valley, Texas." *Waste Management* 28, no. 5 (2008): 776–794. https://doi.org/10.1016/j.wasman.2007.09.013.
32. Genovese, Andrea, Akshay A. Acquaye, Alejandro Figueroa, and S. C. Lenny Koh. "Sustainable Supply Chain Management and the Transition Towards a Circular Economy: Evidence and Some Applications." *Omega* 66 (2017): 344–357. https://doi.org/10.1016/j.omega.2015.05.015.
33. Heidrich, Oliver, Julia Harvey, and Norman Tollin. "Stakeholder Analysis for Industrial Waste Management Systems." *Waste Management & Research* 27, no. 9 (2009): 952–960. https://doi.org/10.1177/0734242X09337679.
34. Mokrini, Abdelilah, Jalal El Alami, and Abderrahim Sayouti. "A Fuzzy Logic Approach for Performance Assessment in Circular Economy." *Procedia Computer Science* 130 (2018): 814–821. https://doi.org/10.1016/j.procs.2018.04.155.
35. Prieto-Sandoval, Vanessa, Carmen Jaca, and Marta Ormazabal. "Towards a Consensus on the Circular Economy." *Journal of Cleaner Production* (2017). https://doi.org/10.1016/j.jclepro.2017.12.224.
36. Merli, Roberto, Michele Preziosi, and Alessia Acampora. "How Do Scholars Approach the Circular Economy? A Systematic Literature Review." *Journal of Cleaner Production* 178 (2018): 703–722. https://doi.org/10.1016/j.jclepro.2017.12.112.

37. Korhonen, Jouni, Cali Nuur, Andreas Feldmann, and Seyoum Eshetu Birkie. "Circular Economy as an Essentially Contested Concept." *Journal of Cleaner Production* 175 (2017): 544–552. https://doi.org/10.1016/j.jclepro.2017.12.111.
38. OECD. "The Circular Economy in Groningen, Netherlands." 2020. https://doi.org/10.1787/e53348d4-en.
39. Norouzi, Masoud, Marta Chàfer, Luisa Cabeza, Laureano Jiménez, and Dieter Boer. "Circular Economy in the Building and Construction Sector: A Scientific Evolution Analysis." *Journal of Building Engineering* 44 (2021): 102704. https://doi.org/10.1016/j.jobe.2021.102704.
40. Bocken, Nancy, Elsa Olivetti, Jonathan M. Cullen, José Potting, and Reid Lifset. "Taking the Circularity to the Next Level: A Special Issue on the Circular Economy." *Journal of Industrial Ecology* 21, no. 3 (2017): 476–482. https://doi.org/10.1111/jiec.12606.
41. Pomponi, Francesco, and Alice Moncaster. "Circular Economy for the Built Environment: A Research Framework." *Journal of Cleaner Production* 143 (2017): 710–718. https://doi.org/10.1016/j.jclepro.2016.12.055.
42. Densley Tingley, Danielle, Jannik Giesekam, and Simone Cooper-Searle. "Applying Circular Economy Principles to Reduce Embodied Carbon." In *Embodied Carbon in Buildings: Measurement, Management, and Mitigation*, edited by Francesco Pomponi, Catherine De Wolf, and Alice Moncaster. Cham: Springer International Publishing, 2018. https://doi.org/10.1007/978-3-319-72796-7_12.
43. Ghisellini, Patrizia, Maddalena Ripa, and Sergio Ulgiati. "Exploring Environmental and Economic Costs and Benefits of a Circular Economy Approach to the Construction and Demolition Sector: A Literature Review." *Journal of Cleaner Production* (2018). https://doi.org/10.1016/j.jclepro.2017.11.207.
44. Haupt, Melanie, Carl Vadenbo, and Stefanie Hellweg. "Do We Have the Right Performance Indicators for the Circular Economy? Insight into the Swiss Waste Management System." *Journal of Industrial Ecology* 21, no. 3 (2017): 615–627. https://doi.org/10.1111/jiec.12506.
45. Leising, Eline, Jaco Quist, and Nancy Bocken. "Circular Economy in the Building Sector: Three Cases and a Collaboration Tool." *Journal of Cleaner Production* 176 (2018): 976–989. https://doi.org/10.1016/j.jclepro.2017.12.010.
46. McKinsey Center for Business and Environment. *The Circular Economy: Moving from Theory to Practice*. McKinsey & Company, 2016. https://www.mckinsey.com/business-functions/sustainability-and-resource-productivity/our-insights/the-circular-economy-moving-from-theory-to-practice. Accessed November 7, 2018.
47. World Business Council for Sustainable Development. "8 Business Cases for the Circular Economy." 2017. https://www.wbcsd.org/Programs/Circular-Economy/Factor-10/Resources/8-Business-Cases-to-the-Circular-Economy. Accessed November 7, 2018.
48. Ellen MacArthur Foundation. "Delivering the Circular Economy: A Toolkit for Policymakers." 2015. https://www.ellenmacarthurfoundation.org/publications/delivering-the-circular-economy-a-toolkit-for-policymakers. Accessed November 22, 2018.
49. arlsberg Group. "Sustainability Report." Carlsberg A/S, 2016. https://carlsberggroup.com/media/10920/2016-sustainability-report.pdf. Accessed November 7, 2018.
50. World Economic Forum. *Towards the Circular Economy: Accelerating the Scale up across Global Supply Chain*. World Economic Forum, 2014. http://www3.weforum.org/docs/WEF_ENV_TowardsCircularEconomy_Report_2014.pdf. Accessed November 7, 2018.
51. European Commission. "Communication from the Commission to the European Parliament, the Council, the European Economic and Social Committee and the Committee of the Regions: Closing the Loop—an EU Action Plan for the Circular Economy." 2015. https://eur-lex.europa.eu/resource.html?uri=cellar:8a8ef5e8-99a0-11e5-b3b7-01aa75ed71a1.0012.02/DOC_1&format=PDF. Accessed April 1, 2019.
52. City of Amsterdam. "Action Plan Sharing Economy." 2016. https://www.slideshare.net/shareNL/amsterdam-actionplan-sharing-economy. Accessed November 11, 2018.

53. Fratini, Chiara Farné, Susse Georg, and Michael Søgaard Jørgensen. "Exploring Circular Economy Imaginaries in European Cities: A Research Agenda for the Governance of Urban Sustainability Transitions." *Journal of Cleaner Production* 228 (2019): 974–989.
54. Circle Economy, FABRIC, TNO, Gemeente Amsterdam. "Circular Amsterdam—a Vision and Action Agenda for the City and the Metropolitan Area." 2015. https://www.circle-economy.com/wp-content/uploads/2016/04/CircularAmsterdam-EN-small-210316.pdf. Accessed November 11, 2018.
55. City of Amsterdam. "Sustainable Amsterdam." 2015. https://www.amsterdam.nl/bestuur-organisatie/organisatie/ruimte-economie/ruimte-duurzaamheid/making-amsterdam/amsterdam-outlines/. Accessed November 11, 2018.
56. U. Mining. "Urban Mining: Concepts, Terminology, Challenges." *Waste Management* 45 (2015): 1–3.
57. Baccini, P., and P. Brunner. *Metabolism of the Anthroposphere: Analysis, Evaluation, Design*. The MIT Press, 2012. ISBN 9780262016650.
58. Lederer, J., D. Laner, J. Fellner, and H. Rechberger. "A Framework for the Evaluation of Anthropogenic Resources Based on Natural Resource Evaluation Concepts." In *Proceedings SUM 2014, 2nd Symposium on Urban Mining*. Bergamo, Italy: IWWG—International Waste Working Group. ISBN: 9788862650311; 016.
59. Material Economics. *The Circular Economy—a Powerful Force for Climate Mitigation*. Stockholm, Sweden: Material Economics, 2018.
60. Circle Economy. *The Circularity Gap Report*. Amsterdam, The Netherlands: Circle Economy, 2021.
61. Bressanelli, G., F. Adrodegari, M. Perona, and N. Saccani. "The Role of Digital Technologies to Overcome Circular Economy Challenges in PSS Business Models: An Exploratory Case Study." *Procedia CIRP* 73 (2018): 216–221.
62. Antikainen, M., T. Uusitalo, and P. Kivikytö-Reponen. "Digitalisation as an Enabler of Circular Economy." *Procedia CIRP* 73 (2018): 45–49.
63. Demestichas, K., and E. Daskalakis. "Information and Communication Technology Solutions for the Circular Economy." *Sustainability* 12 (2020): 7272.
64. Rosa, P., C. Sassanelli, A. Urbinati, D. Chiaroni, and S. Terzi. "Assessing Relations between Circular Economy and Industry 4.0: A Systematic Literature Review." *International Journal of Production Research* 58 (2019): 1662–1687.
65. Wilts, H., and H. Berg. *The Digital Circular Economy: Can the Digital Transformation Pave the Way for Resource-Efficient Materials Cycles?* Wuppertal, Germany: Wuppertal Institute, 2017.
66. Climate-KIC. *Digitalisation—Unlocking the Potential of the Circular Economy*. White Paper. Brussels, Belgium: Climate-KIC, 2018.
67. Kristoffersen, E., F. Blomsma, P. Mikalef, and J. Li. "The Smart Circular Economy: A Digital-Enabled Circular Strategies Framework for Manufacturing Companies." *Journal of Business Research* 120 (2020): 241–261.
68. Riesener, M., C. Dolle, C. Mattern, and J. Kres. "Circular Economy: Challenges and Potentials for the Manufacturing Industry by Digital Transformation." In *Proceedings of the 2019 IEEE International Symposium on Innovation and Entrepreneurship (TEMS-ISIE)*, Hangzhou, China, October 24–26, 2019, pp. 1–7.
69. European Commission. "Europe's Digital Transition Goes Hand in Hand with the European Green Deal." June 7, 2022. https://digital-strategy.ec.europa.eu/en/policies/green-digital.
70. Bohnsack, R., C. M. Bidmon, and J. Pinkse. "Sustainability in the Digital Age: Intended and Unintended Consequences of Digital Technologies for Sustainable Development." *Business Strategy and the Environment* 31, no. 2 (2022): 599–602.
71. Blackburn, O., P. Ritala, and J. Keränen. "Digital Platforms for the Circular Economy: Exploring Metaorganizational Orchestration Mechanisms." *Organization & Environment*. Advance online publication (2022).

72. IKEA. "Designing for Circularity: IKEA's Circular Design Principles." 2020. https://www.ikea.com/gb/en/this-is-ikea/sustainable-everyday/designing-for-circularity-pub6b0e09a5.
73. IKEA. "Giving Furniture a Second Life." 2019. https://www.ikea.com/ms/en_US/this-is-ikea/people-and-planet/giving-furniture-a-second-life/index.html.
74. IKEA. "Innovative Materials for a Sustainable Future." 2021. https://www.ikea.com/us/en/this-is-ikea/sustainable-everyday/innovative-materials-for-a-sustainable-future-pub6f4e09a6.
75. Ellen MacArthur Foundation. "IKEA's Circular Value Chain." 2020. https://www.ellenmacarthurfoundation.org/case-studies/ikeas-circular-value-chain.
76. Cambridge Econometrics, Trinomics, and ICF. *Impacts of Circular Economy Policies on the Labour Market*. Luxembourg: Publications Office of European Union, 2018. http://publications.europa.eu/publication/manifestation_identifier/PUB_KH0418564ENN.
77. Sepúlveda-Rojas, J. P., and P. A. Benitez-Fuentes. "Coordination and Return Uncertainties in Closed Loop Supply Chains." In *Paper Presented at the 2016 6th International Conference on Computers Communications and Control (ICCCC)*, pp. 188–195. https://doi.org/10.1109/ICCCC.2016.7496759.
78. Hvass, K. K., and E. R. G. Pedersen. "Toward Circular Economy of Fashion: Experiences from a Brand's Product Take-Back Initiative." *Journal of Fashion Marketing and Management: An International Journal* (2019). https://doi.org/10.1108/JFMM-04-2018-0059.
79. International Resource Panel (IRP). *Re-Defining Value—the Manufacturing Revolution. Remanufacturing, Refurbishment, Repair and Direct Reuse in the Circular Economy*. A Report of the International Resource Panel. Nairobi, Kenya: United Nations Environment Programme, 2018.
80. Russell, Jennifer D., and Nabil Z. Nasr. "Value-Retained vs. Impacts Avoided: The Differentiated Contributions of Remanufacturing, Refurbishment, Repair, and Reuse within a Circular Economy." *Journal of Remanufacturing* 13, no. 1 (2023): 25–51.
81. Reike, D., W. J. Vermeulen, and S. Witjes. "The Circular Economy: New or Refurbished as CE 3.0?—Exploring Controversies in the Conceptualization of the Circular Economy through a Focus on History and Resource Value Retention Options." *Resources, Conservation and Recycling* 135 (2018): 246–264.
82. Tukker, A. "Product Services for a Resource-Efficient and Circular Economy–a Review." *Journal of Cleaner Production* 97 (2015): 76–91.
83. Oskamp, Stuart. "Resource Conservation and Recycling: Behavior and Policy." *Journal of Social Issues* 51, no. 4 (1995): 157–177.
84. Prochatzki, Georg, Ralph Mayer, Josephin Haenel, Anja Schmidt, Uwe Götze, Martin Ulber, Anne Fischer, and Marlen Gabriele Arnold. "A Critical Review of the Current State of Circular Economy in the Automotive Sector." *Journal of Cleaner Production* (2023): 138787.
85. Bakker, C., F. Wang, J. Huisman, and M. den Hollander. "Products that Go Round: Exploring Product Life Extension Through Design." *Journal of Cleaner Production* 69, no. 1 (2014): 10–16.
86. Schroeder, P., K. Anggraeni, and U. Weber. "The Relevance of Circular Economy Practices to Sustainable Development Goals." *Journal of Industrial Ecology* 23, no. 1 (2019): 77–95.
87. Manavalan, E., and K. Jayakrishna. "An Analysis on Sustainable Supply Chain for Circular Economy." *Procedia Manufacturing* 33, no. 1 (2019): 477–484.
88. Ellen MacArthur Foundation. "Remanufacturing at Renault." Accessed August 7, 2023. https://www.ellenmacarthurfoundation.org/case-studies/remanufacturing-at-renault.
89. Renault Group. "Remanufacturing: Giving Parts a Second Life." Accessed August 7, 2023. https://www.renaultgroup.com/en/news-on-air/news/renault-remanufacturing-giving-parts-a-second-life/.

90. Stahel, W. R. "The Circular Economy." *Nature* 531, no. 7595 (2016): 435–438. https://doi.org/10.1038/531435a.
91. Ceder, G. "Accelerating Materials Discovery with High-Throughput Methods." *Nature Materials* 16, no. 2 (2017): 101–108.
92. Statista. "Consumer Electronics Report 2017—Telephony." Statista Consumer Market Outlook—Segment Report. http://tinyurl.com/y6kotd7e. Accessed August 7, 2023.
93. Ellen MacArthur Foundation. "Circular Consumer Electronics: An Initial Exploration." 2018. http://tinyurl.com/yycsm7ud. Accessed August 7, 2023.
94. Lee, P., D. Stewart, and C. Calugar-Pop. "Technology, Media and Telecommunications Predictions 2018." Deloitte, 2017. Accessed August 7, 2023. https://tinyurl.com/y5j2gao5.
95. European Commission. *Impact of the Structural Changes on Jobs and Industrial Relations in the Telecommunications and ICT: Final Report*. Publications Office of the European Union, 2015.
96. Verbeek, L. H. "A Circular Economy Index for the Consumer Goods Sector." Master's thesis, 2016.
97. Dell Technologies. "Progress Made Real: Annual Report on 2030 Goals." https://www.dell.com/learn/us/en/uscorp1/progress-made-real-report. Accessed August 7, 2023.
98. Apple Inc. "Environmental Progress Report." Accessed August 7, 2024. https://www.apple.com/environment/pdf/Apple_Environmental_Progress_Report_2024.pdf.
99. Ellen MacArthur Foundation. "Circular Economy Case Studies." Accessed August 7, 2023. https://www.ellenmacarthurfoundation.org/case-studies.
100. HP Inc. "Sustainability Report." Accessed August 7, 2024. https://www.hp.com/us-en/hp-information/sustainable-impact.html.
101. Brundtland, G. H. *Our Common Future*. Oxford: Oxford University Press, 1987.
102. United Nations. "Transforming Our World: The 2030 Agenda for Sustainable Development." Accessed August 7, 2015. https://sustainabledevelopment.un.org/post2015/transformingourworld/publication.
103. Philips. "Sustainability Report." Accessed August 7, 2023. https://www.philips.com/a-w/about/sustainability.html.
104. Interface. "Mission Zero Progress Report." Accessed August 7, 2024. https://www.interface.com/US/en-US/about/mission-zero.
105. BMW Group. "Sustainable Value Report." Accessed August 7, 2023. https://www.bmwgroup.com/en/responsibility/sustainability.html.
106. Bahrin, Mohd Aiman Kamarul, Mohd Fauzi Othman, Nor Hayati Nor Azli, and Muhamad Farihin Talib. "Industry 4.0: A Review on Industrial Automation and Robotic." *Jurnal Teknologi* 78, nos. 6–13 (2016).
107. International Federation of Robotics. "World Robotics Report." 2022. https://ifr.org/world-robotics-report-2022. Accessed August 7, 2023.
108. Robotics Business Review. "Robots in Oil & Gas: Driving Efficiency and Safety." 2021. https://www.roboticsbusinessreview.com/robots-in-oil-gas-driving-efficiency-and-safety/. Accessed August 7, 2023.
109. McKinsey & Company. "The Future of Robotics in Quality Control." 2020. https://www.mckinsey.com/business-functions/operations/our-insights/the-future-of-robotics-in-quality-control. Accessed August 7, 2023.
110. Geim, A. K., and K. S. Novoselov. "The Rise of Graphene." *Nature Materials* 6, no. 3 (2007): 183–191. https://www.nature.com/articles/nmat1849.
111. Tokiwa, Y., and B. P. Calabia. "Biodegradability and Biodegradation of Poly(lactide)." *Applied Microbiology and Biotechnology* 72, no. 2 (2006): 244–251. https://link.springer.com/article/10.1007/s00253-006-0488-1.
112. Chen, Q., and G. A. Thouas. "Metallic Implant Biomaterials." *Materials Science and Engineering: R: Reports* 87 (2015): 1–57. https://www.sciencedirect.com/science/article/pii/S0927796X14000793.

113. Butler, K. T., D. W. Davies, H. Cartwright, O. Isayev, and A. Walsh. "Machine Learning for Molecular and Materials Science." *Nature* 559, no. 7715 (2018): 547–555.
114. Harrop, P. "AI and Robotics for Accelerated Material Development." *Materials Today* 22, no. 3 (2019): 148–153.
115. Mamalis, A. "Recent Advances in Nanotechnology." *Journal of Materials Processing Technology* 181, nos. 1–3 (2007): 52–58.
116. Scott, T. O., D. R. Ewim, and A. C. Eloka-Eboka. "Hybrid Nanofluids Flow and Heat Transfer in Cavities: A Technological Review." *International Journal of Low-Carbon Technologies* ctac093 (2022).
117. Lee, Phillip, Jinhwan Lee, Hyungman Lee, Junyeob Yeo, Sukjoon Hong, Koo Hyun Nam, Dongjin Lee, Seung Seob Lee, and Seung Hwan Ko. "Highly Stretchable and Highly Conductive Metal Electrode by Very Long Metal Nanowire Percolation Network." *Advanced Materials* 24, no. 25 (2012): 3326–3332.
118. Giri, P., D. Goswami, and A. Perumal. "Advanced Nanomaterials and Nanotechnology." Paper presented at the *Proceedings of the 2nd International Conference on Advanced Nanomaterials and Nanotechnology*, 2013.
119. Baig, N., I. Kammakakam, and W. Falath. "Nanomaterials: A Review of Synthesis Methods, Properties, Recent Progress, and Challenges." *Materials Advances* 2, no. 6 (2021): 1821–1871.
120. Dubois, D., and H. Prade. "A Survey of Fuzzy Logic Applications in Environmental and Resource Management." *Environmental Modelling & Software* 37 (2012): 1–13.
121. Karwowski, W., and G. W. Evans. "Fuzzy Logic in Sustainable Waste Management." *Journal of Environmental Management* 242 (2019): 489–498.
122. Govindan, K., and M. Bouzon. "From a Literature Review to a Multi-Perspective Framework for Circular Supply Chains." *Journal of Cleaner Production* 197 (2018): 530–541.
123. Zadeh, Lotfi A. "Fuzzy Sets." *Information and Control* 8, no. 3 (1965): 338–353.
124. Klir, George J., and Bo Yuan. *Fuzzy Sets and Fuzzy Logic: Theory and Applications*. Prentice Hall, 1995.
125. Kosko, Bart. *Fuzzy Thinking: The New Science of Fuzzy Logic*. Hyperion, 1994.
126. Arayici, Yavuz, and Peter Coates. "A System for Modelling Waste Management Processes." *Waste Management Journal* 32, no. 3 (2012): 431–439.
127. Huisman, Jaco, Christian Meskers, Xianlai Zeng, Maria V. D. P. Salgado, and Feng Wang. "The E-Waste Challenge: Optimizing Resource Recovery Pathways." *Journal of Cleaner Production* 252 (2020): 119736.
128. Tittonell, Pablo, and Keith E. Giller. "When Yield Gaps Are Poverty Traps: The Paradigm of Ecological Intensification in African Smallholder Agriculture." *Field Crops Research* 143 (2013): 76–90.
129. Angelidaki, Irini, and Lars Ellegaard. "Codigestion of Manure and Organic Wastes in Centralized Biogas Plants: Status and Future Trends." *Applied Biochemistry and Biotechnology* 109, nos. 1–3 (2003): 95–105.
130. Baky, Ahmed, and Olof Eriksson. "Systems Analysis of Organic Waste Management in Denmark." *Resources, Conservation and Recycling* 38, no. 1 (2003): 89–102.
131. Gebreslassie, Berhane H., Mark Slivinsky, Bing Wang, and Feng You. "Life Cycle Optimization for Sustainable Design and Operations of Hydrocarbon Biorefinery via Fast Pyrolysis, Hydrotreating and Hydrocracking." *Computers & Chemical Engineering* 50 (2013): 71–91.
132. Shen, Lihui, Jürgen Haufe, and M. K. Patel. "Product Overview and Market Projection of Emerging Bio-Based Plastics." *PRO-BIP 2009* 243 (2009): 61–69.
133. Van Selm, Benjamin, Anita Frehner, Imke JM De Boer, Ollie Van Hal, Renske Hijbeek, Martin K. Van Ittersum, Elise F. Talsma et al. "Circularity in animal production requires a change in the EAT-Lancet diet in Europe." *Nature Food* 3, no. 1 (2022): 66–73.

134. Geissdoerfer, Martin, Paulo Savaget, Nancy M. Bocken, and Eric J. Hultink. "The Circular Economy—a New Sustainability Paradigm?" *Journal of Cleaner Production* 143 (2017): 757–768.
135. Antikainen, Maija, and Katriina Valkokari. "A Framework for Sustainable Circular Business Model Innovation." *Technology Innovation Management Review* 6, no. 7 (2016): 5–12.
136. Carus, Michael, Lars Dammer, and Richard Essel. "Options for Designing a New Political Framework of the European Bio-Based Economy." *Industrial Biotechnology* 10, no. 2 (2014): 69–71.
137. European Commission. "Circular Economy Action Plan: For a Cleaner and More Competitive Europe." 2020. Accessed August 7, 2024. https://ec.europa.eu/environment/circular-economy/.
138. Kirchherr, Julian, Laura Piscicelli, Robin Bour, Eline Kostense-Smit, Jochem Muller, Anke Huibrechtse-Truijens, and Marko Hekkert. "Barriers to the Circular Economy: Evidence from the European Union (EU)." *Ecological Economics* 150 (2018): 264–272.
139. Bhardwaj, A., S.V. Kumar, A. Vinukonda, K. Ramakrishna, S. Srivastava, S.P. Rajendran, and P. Sharma. "Robotics and Automation: Revolutionizing Research and Development." In *Evolving Landscapes of Research and Development: Trends, Challenges, and Opportunities 2025* (pp. 313–338). IGI Global Scientific Publishing.

12 Challenges and Future Directions

S. Balakrishnan, Sunil Kumar RM, and S. Simonthomas

12.1 CHALLENGES IN ENGINEERING THE CIRCULAR ECONOMY

The transition to a circular economy is a very challenging task that demands a great deal of change in design, production, and consumption methods [1]. The following analysis outlines the major challenges that must be addressed in the engineering of a circular economy:

1. Processes for material recovery and recycling
2. Product design for circularity
3. Integration of renewable energy frameworks
4. Supply chain complexity in circular systems
5. Viability of economic models and stakeholder involvement
6. Policy and regulatory hurdles
7. Technological deficiencies and constraints

12.1.1 Processes for Material Recovery and Recycling

Heterogeneity of materials: Most products involve multiple material combinations that prove impossible to separate and recover during recycling. *Degradation of material quality:* Recycling is repetitive in nature, which reduces material quality and limits its reuse in the most critical applications. *Infrastructure issues:* Poor recycling facilities and disorganized waste collection systems limit high recovery rates of materials. *Economic barriers:* Generally, the financial outlay for recycling activities is greater than that for the extraction of raw materials, thereby deterring their adoption [2].

12.1.2 Product Design for Circularity

Lack of standardization: Product design for circularity needs standardization of materials and components, which is not yet practiced. *Design trade-offs:* Achieving a balance among durability, recyclability, and functionality in product design presents considerable challenges. *End-of-life considerations:* Engineers frequently encounter difficulties in integrating strategies for straightforward disassembly, repair, and recycling within their product designs.

DOI: 10.1201/9781003538837-12

12.1.3 Integration of Renewable Energy Frameworks

Energy demand inequality: Since circular economy activities, such as recycling and remanufacturing, are energy-intensive, not coming from renewable sources, this will create an inequity in energy. *Technological inadequacy:* The absence of breakthrough advancement in energy storage technology and, more importantly, the integration of renewable energy technologies, hampers the transition. *Transition expenses:* The considerable initial capital required for renewable energy infrastructure discourages enterprises from embracing circular energy models.

12.1.4 Supply Chain Complexity in Circular Systems

Reverse logistics has to do with the detailed and sometimes costly management of returns concerning products and materials for recyclable or reusable purposes. The establishment of circular systems requires collaboration between supply chains that are distributed geographically, which complicates efforts at coordination. *Lack of transparency:* There is no adequate information that is related to material flow along with different product life cycle stages, which prevents efficient administration of circular supply chains [3].

12.1.5 Viability of Economic Models and Stakeholder Involvement

Substantial upfront investments: Circularity transition requires a lot of financial inputs in technology and infrastructure investment. *Unstable market demand:* In addition, the demand for the product is bound to vary from customer's side to another with respect to remanufactured or recycled products. *Luddites and transformation:* Implementing this linear to circular change of paradigm will always be tough and needs overcoming conventional behavior of the stakeholders.

12.1.6 Policy and Regulatory Hurdles

Inconsistent policies: Different policies in various regions challenge businesses that work in different jurisdictions. *Limited government support:* Strong government incentives in supporting businesses are low, which discourages firms from embracing circularness. Ineffective enforcement of environmental regulations can cause much difficulty in the success of circular economy ventures.

12.1.7 Technological Deficiencies and Constraints

Lack of Technological Leap: Many circular economy initiatives rely on the technologies which are still developing and not yet financially ready. *Data and analytics:* The use of advanced analytical tools, such as artificial intelligence and the Internet of Things, for the resource usage tracking and optimization is very minimal, thus restricting progress. *The main challenges in scalability:* Circular technologies prototypes often fail to scale up to industrial levels due mainly to issues concerning cost and complexity.

Combating the problems necessitates an integrated approach toward technology development, policy adjustment, and cultural evolution of behavior in both production and consumption. The overcoming of these challenges will be critical to the successful development of a circular economy.

12.2 CASE STUDIES AND APPLICATIONS

Establishing a circular economy calls for novel strategies and technologies in an attempt to rethinking systems, increasing resource use efficiency, and reducing waste. Some examples and case studies follow for a comprehensive overview of how the different sectors have adopted the principles of the circular economy [4].

12.2.1 Fuzzy Logic in Sustainable Manufacturing

- *Case study:* Resource allocation in a furniture manufacturing plant
- *Problem:* Inefficiencies in wood usage were causing huge waste.
- *Solution:* Fuzzy logic was used to improve the cut patterns of timber logs, so less waste material was produced and more usable output material was realized.
- *Outcome:* Overall, 25% reduction in material waste and 15% improvement in production efficiency.
- *Application:* The use of fuzzy algorithms for dynamic energy management in production facilities aligns the consumption of energy with renewable sources, which eventually decreases dependence on fossil fuels.

12.2.2 Circular Supply Chains

- *Case study:* Nike's reuse-a-shoe program
- *Problem:* Managing postconsumer waste from discarded athletic shoes.
- *Solution:* Nike created a reverse logistics system to collect the used shoes, break down the materials into raw form, and then use them to refurbish playground surfaces and sports fields.
- *Outcome:* More than 30 million pairs of shoes processed which reduced landfill waste and promoted healthy practices.
- *Application:* Using IoT with blockchain to track the whole life cycle of products and ensure proper recycling when their life cycle ends.

12.2.3 Product Design for Circularity

- *Case study:* Fairphone—modular smartphone design
- *Problem:* Traditional smartphones have a limited lifespan because they are not repairable.
- *Solution:* Fairphone has used modular design, where customers can change specific parts of the phone like a battery or screen instead of having to dispose of the whole device.

- *Outcome:* Increased product lifetime, minimized e-waste, and increased consumer interaction with circularity.
- *Application:* Applying 3D printing to make modular components on demand results in very low risks of overproduction and waste.

12.2.4 Waste-to-Resource Conversion

- *Case study:* Veolia's wastewater treatment in Denmark
- *Problem:* The treatment of industrial wastewater is very costly and energy-intensive.
- *Solution:* Veolia implemented anaerobic digestion in order to exploit biogas from sludge waste generated from wastewater. It is then transformed into a source of renewable energy.
- *Outcome:* The plant became energy-neutral, recovering 150% of its energy consumption from biogas production.
- *Application:* Determine with fuzzy logic the optimum conditions for anaerobic digestion that optimize biogas output under a given energy use restriction.

12.2.5 Industrial Symbiosis

- *Case study:* Kalundborg Symbiosis, Denmark
- *Problem:* While wastes from one industry were being discharged, nearby industries faced shortages.
- *Solution:* Kalundborg city developed a collaborative framework in which waste by-products, such as steam, heat, and water, produced by one organization were efficiently used as inputs by another organization. For example, excess steam from a power plant was used by a pharmaceutical company and a fish farm.
- *Outcome:* Substantial cost-savings, reduced emissions, and increased resource efficiency across the board in different industries.
- *Application:* AI-based system for the possible synergy identification between various industries and for optimizing material flow.

12.2.6 Renewable Energy Integration

- *Case study:* Tesla's battery recycling program
- *Problem:* Managing end-of-life electrical vehicle batteries and renewable energy storage systems at the end.
- *Solution:* Using its leading-edge recycling technology, Tesla has successfully reclaimed lithium, cobalt, and nickel from old batteries and has fed these into new cycles of production.
- *Outcome:* Reduced dependency on mining raw materials and closed the loop for battery materials.
- *Application:* Development of smart grids based on renewable energy, and integration of fuzzy logic systems to optimize energy storage and distribution according to prevailing demand conditions.

12.2.7 Food Waste Management

- *Case study:* Closed-loop food packaging
- *Problem:* Single-use food packaging contributes significantly to waste.
- *Solution:* Loop teamed up with leading food manufacturers to launch hardy, returnable containers for ice cream and other beverages.
- *Outcome:* Encourage consumers to bring in their containers to be cleaned and reused to decrease the level of packaging waste and sensitize the consumer on the circular economy.
- *Application:* The technology in use is AI-improved sorting facilities that ensure efficient segregation of organic wastes to composting or biogas production.

12.2.8 Circular Economy in the Construction Sector

- *Case study:* Waste-to-material recycling in Amsterdam
- *Problem:* Building and demolition waste accounted for high landfill utilization.
- *Solution:* Amsterdam developed a concrete, wood, and metals recycling program from the old building to construct in their new building.
- *Outcome:* More than 80% of Amsterdam construction waste is recycled today, thus saving virgin material content.
- *Application:* Fuzzy decision-making systems to determine the acceptability of recycled materials to be used in construction.

12.2.9 Circular Fashion and Textiles

- Case study: Patagonia's worn wear initiative
- *Problem:* The fast fashion industry produces large amounts of textile waste.
- *Solution:* Patagonia encouraged customers to return the used clothes for repair or resale or to be reformed into new clothes.
- *Outcome:* Increasing life cycle of clothing usage and increasing consumer consciousness about sustainable use of clothing.
- *Application:* Development of AI and fuzzy logic models for grading returns of textiles to be repaired, resold, or recycled.

12.2.10 Circular Economy in Electronics

- *Case study:* Dell's closed-loop recycling
- *Problem:* Valuable materials are dumped in electronic waste.
- *Solution:* Dell has an invention of a closed system aiming to recover plastics and metal of old computers, which afterward utilizes them in the manufacturing process.
- *Outcome:* 95% of materials from recycled devices are now reused, thereby reducing the extraction of resources.
- *Application:* Automated disassembly of electronic devices using robots backed up by fuzzy logic for optimal recovery of parts.

The case studies and applications underline the flexibility and effectiveness of the engineering solutions in promoting the circular economy [5]. By showing the design of modular products, waste-to-resource conversion, renewable energy systems, and industrial symbiosis, such initiatives prove that technological innovation, together with collaborative initiatives, can propel the movement to a sustainable and circular future.

12.3 FUTURE INNOVATIONS AND DIRECTIONS

The future of engineering the circular economy, especially when applying fuzzy logic, promises many directions and innovative strategies [6].

12.3.1 Future Innovations

The following are some key areas of focus and development.

12.3.1.1 Integration of Advanced Digital Technologies

- *AI and IoT synergies:* Fuzzy logic will then be integrated with AI and the Internet of Things (IoT) to enhance decision-making in circular systems. This approach allows the dynamic monitoring and improvement of resource movement in manufacturing, waste management, and supply chain [7].
- *Blockchain for transparency:* Blockchain technology enables traceability in circular processes and fosters better resource tracking for consumer trust in sustainable product.

12.3.1.2 Innovations in Design and Manufacturing

- *Smart product design:* The modular and sustainable design principles make easy repairability, upgradability, or recyclability of the products. For the enhancement of designs toward sustainability and efficient resource use, the simulations using fuzzy logic may come into play.
- *Sustainable material use:* Biomaterial and recycled composite innovations may be guided by fuzzy logic systems to balance the costs of functionality and environmental impact.

12.3.1.3 Sector-Specific Advancements

- *Construction:* Application of fuzzy logic can control waste and promote recycling during the construction phase. Alternative cement materials and recycled aggregates are examples of innovation in the extension of the lifetime of construction resources [8].
- *Textiles:* In textiles, circular approaches with the support of fuzzy decision-making can make it possible to improve the reuse and recycling of fabrics and thus decrease both water usage and emissions.

12.3.1.4 Policy and Market Mechanisms

- *Circular economy policies:* Government use fuzzy logic to simulate effects of circular policies, say taxes on waste or incentives on recycled materials.

These resources provide critical information about the results associated with the economy and the environment.

- *Market innovations:* New business models, such as product-as-a-service, facilitated by fuzzy-based analytics, can be used to optimize better resource allocations and life cycle management of products.

12.3.1.5 Collaborative Platforms

- *Cross-industry collaborations:* Programs such as Scale 360° of the World Economic Forum help promote cross-industry adoption of circular practice through full toolkits and methodologies intended to enhance the scaling of innovations.
- *Educational and training programs:* Knowing these principles of circular economy, if added with fuzzy logic applications, transition can be highly accelerated in virtually all industries.

12.3.2 Future Research Directions

- *AI-fuzzy hybrid models:* Combine fuzzy logic and machine learning procedures to improve the performability of circular economy frameworks.
- *Dynamic systems modeling:* Dynamic systems modeling is a process aimed at designing predictive tools that predict long-term effects of circular interventions in the presence of uncertainty.
- *Scalable solutions:* A design of adaptable frameworks by the circular economy for small and medium enterprises (SMEs).

These developments have brought into the fore the great importance of fuzzy logic in implementing the principles of a circular economy, which would eventually support economic development, resource efficiency, and environmental sustainability.

12.3.2.1 Advances in Fuzzy Logic Technologies

Moving from a linear economic system to a circular economic model involves multidimensional choices over resource use efficiency, minimization of waste, and the optimization of sustainable practices. Such a change would call for a different style of decision-making compared with traditional ones and would need very robust techniques that can cope with uncertainty, imprecision, and complex interdependencies. Fuzzy logic has proven to be such a robust tool because it originated as a subset of artificial intelligence [9].

(i) *Key Applications of Fuzzy Logic in Circular Economy*

1. Resource optimization fuzzy logic systems enable accurate decisions in respect of resource allocation, recycling of materials, and usage of energy. For instance, fuzzy controllers improve renewable energy use in manufacturing process leading to reduced waste and increase productivity [10].
2. Waste management advanced fuzzy models in waste management classify, separate, and prioritize materials by looking at the waste's impact on the environment, recyclability potential, and the monetary value

associated with the same. This way, sustainable as well as financially effective innovative waste management systems could be developed.

3. Sustainable supply chain management fuzzy logic improves the supply chain decision by taking into account many more sustainability criteria, such as carbon footprint, material lifetime, and logistics efficiency. The system creates closed-loop operations as well as reverse logistics [11].
4. Eco-design and life cycle assessment (LCA) engineers use fuzzy systems to assess the environmental impact of products over their entire life cycle. Integration of fuzzy logic with the life cycle assessment approaches gives designers the chance to produce sustainable products that will have long lifetimes, recyclable materials, and minimal adverse impacts on the environment.
5. Policymakers rely on fuzzy logic to reconcile environment, social, and economic objectives. Fuzzy multi-criteria decision-making (MCDM) models are very important for designing CE policies that can respond to the multiple needs of the stakeholders and the uncertainties inherent in those needs.

(ii) Recent Advances

- Real-time monitoring and adaptive control are possible due to the integration of fuzzy systems with IoT sensors and big data analytics in operations like the circular economy-orientated smart grids and waste-to-energy facilities.
- Hybrid models: The hybrid models which are an integration of fuzzy logic, machine learning, and genetic algorithms will make CE solutions more accurate and robust.
- Cloud-based fuzzy systems: These facilitate scalable and distributed decision-making for global CE initiatives, and facilitate industries, governments, and communities in collaborating.

(iii) Impact on Engineering Practices

Fuzzy logic technologies redefine engineering approaches in CE by pushing data-driven, adaptive, and inclusive solutions. They lead to innovation in materials, in energy systems, in urban development, and in industrial process engineering practices and bring those practices into the context of sustainable development goals.

A resilient and sustainable circular economy will be accelerated with the rapid move of industries and governments embracing these new advances.

12.3.2.2 AI and Machine Learning Synergies with Fuzzy Logic

The concept of a circular economy hopes to introduce systems that are resource-conservative in nature by means of reorienting designing, utilizing, and recycling resources-based methodologies. In its pursuit, massive and highly complex datasets come into the scope of management activities and dealing with uncertainty, while all this forms fertile fields for artificial intelligence, machine learning and fuzzy logic approaches. And the integration of these approaches results in groundbreaking applications for the conception of a circular economy [12].

(i) *Key Synergies between AI, ML, and Fuzzy Logic in CE*

1. **Optimized Decisions in Noisy Environments**

   More reliable decisions under uncertainty fuzzy logic represent a robust framework to tackle uncertainty and vagueness commonly met in CE contexts, like the forecasting of material lifetimes or assessing sustainability indicators. Machine learning complements this by training models on historical data, improving the precision and adaptability of fuzzy systems.

   For instance, integration of AI into a fuzzy system can enhance recycling activities through control variables such as energy efficiency and material sorting in an information-based environment.

2. **Predictive Analytics for Resource Management**

   ML algorithms go through big data to understand the pattern, like how fast materials degrade or consume trends. Fuzzy logic analyzes these outputs to give actionable predictions, hence allowing resources to be used efficiently.

   Example: Smart waste systems use ML for waste volume prediction and fuzzy logic to prioritize recycling routes and method.

3. **Optimization of Closed-Loop Systems**

   For closed-loop supply chains, artificial intelligence and machine learning are critical for the analysis of data to identify inefficiencies; for fuzzy logic, there are evaluations of trade-offs of different objectives such as cost- and energy-efficiency or recyclability.

   For example, using AI-based fuzzy systems for reverse logistics might optimize it through route proposal that is economically viable, but also friendly to the environment for the return of used products.

4. **Integration with IoT for Real-Time Monitoring**

   Internet of Things devices, in the context of circular economy applications, generate copious data on energy consumption, waste monitoring, and efficiency of production. The data goes through ML, but it is the fuzzy logic that helps to create decision-making systems with real-time adaptability in circular processes.

   For instance, this technology has a very interesting application in smart grids that apply AI-fuzzy hybrids in order to manage the balance between supply and demand of renewable energy.

5. **Life Cycle Assessment and Eco-Design**

   ML models will be used to identify areas of significant environmental impacts along the entire life cycle of a product; fuzzy logic improves multi-criteria decision-making to design sustainable products.

   Example: AI-fuzzy systems propose alternative materials or manufacturing processes in accordance with sustainability targets.

(ii) *Recent Innovations in AI-Fuzzy Synergies for CE*

Neuro-fuzzy systems: The system combines the learning of neural networks and the interpretative power of fuzzy logic to improve decision-making in waste sorting and energy optimization in the circular economy.

Reinforcement learning with fuzzy logic: In this approach, reinforcement learning can be applied to train agents in circular processes optimization by time. Meanwhile, rules of fuzzy logic ensure decision interpretability and conformity with the sustainability objective.

Explainable AI (XAI): The integration of fuzzy logic with the XAI will make AI decisions about CE applications more transparent, helping to boost trust and regulatory compliance.

(iii) Circular Economy Engineering Impact

1. *Optimization of sustainability:* With AI–ML–fuzzy systems, engineers can design adaptive, efficient systems for energy recovery, material reuse, and waste reduction.
2. *Cost-effective operations:* Predictive maintenance needs along with optimization of resource flows in these technologies reduce operational costs while enhancing sustainability.
3. *Scalable solutions:* Hybrid AI-fuzzy models scale seamlessly, allowing global industries to implement CE principles across supply chains and manufacturing ecosystems.

(iv) Future

The interaction of AI, ML, and fuzzy logic will push the innovation of engineering in the circular economy. The future may include autonomous systems for CE powered by real-time AI-fuzzy models, smarter predictive maintenance for industrial assets, and fully integrated digital twins of circular processes.

With these technologies, industries can hasten the journey to a sustainable future where waste becomes a resource and efficiency is the norm.

12.3.2.3 Smart Cities and Circular Economy Integration

The integration of smart cities and the circular economy is one of the transformational approaches to sustainable urban development. Integrating advanced technologies with the circular economy principle allows urban spaces to enhance resource efficiency, waste reduction, and the growth of robust ecosystems that benefit both the economy, environment, and society.

(i) Key Principles of Integration

1. **Resource Efficiency**
   High-tech urban regions apply Internet of Things, artificial intelligence, and big data for real-time resource monitoring and management—from energy to water to raw materials. These are used under circular economy principles and therefore minimized or reduced to be wasted.
2. **Waste as a Resource**
   The circular economy framework in the smart city context perceives waste as an innovation potential. It becomes possible to increase tracking, sorting, and processing of wastes using some advanced technologies in a way such that better recycling and energy recovery will be attainable.

3. **Decentralized Systems**
   Smart cities are always decentralized for managing energy, water, and waste to control resources at local levels and remove inefficiencies. Circular practices introduced to these systems optimize them through closed loops by using treated wastewater to irrigate land and energy to come from organic waste.
4. **Stakeholder Collaboration**
   Smart city initiatives stimulate citizen engagement and encourage cooperation among stakeholders. Circular economy principles therefore enable the collaboration of people in solution creation processes, including community repair hubs and shared mobility systems.

(ii) Technological Enablers for Integration

1. **Internet of Things (IoT)**
   Sensors of IoT can sense in real-time consumption, waste generation, and emission of resources. An example will be smart bin fitted with sensors tracking waste levels in a neighborhood and optimizing routes of waste collection for more recycling.
2. **Artificial Intelligence (AI) and Big Data**
   Artificial intelligence analyzes vast data collections to identify inefficiencies and predict the need for resources. Predictive models help manage energy use while machine learning further improves circular supply chain material flow optimization.
3. **Blockchain**
   Blockchain ensures transparency and traceability in circular systems—when it follows the recycling flows within and across supply chains. *Example:* Digital passports for products are necessary to enable good repair, reuse, and recycling through detailed life cycle information.
4. **Digital Twins**
   Digital twins present a kind of simulation for urban systems, which is helpful to evaluate circular economy strategies while enhancing resource distribution. Thus, a digital twin based on the city's energy grid can find opportunities related to renewable energy source integration, as well as enhance general efficiency.

(iii) Key Areas of Integration

1. **Smart Energy Grids**
   Adoption of renewable energy is enabled through circular practices and the balancing of supply and demand by smart grids. *Example:* Utilization of waste-to-energy plants for supplementary renewable sources in urban grids.
2. **Water Management**
   Smart sensors track the usage of water and leakage while CE systems reuse treated wastewater for agriculture or industrial purposes. *Example:* The water systems in Singapore are designed as a circle that maximizes reuse in reducing the importation of water.

3. **Sustainable Mobility**
   Smart cities promote shared and electric mobility solutions that fit CE priorities of minimizing emissions and resources. *Example:* Shared modular, recyclable bike services reduce material waste.
4. **Urban Waste Management**
   Advanced sorting technologies, in combination with AI systems, upgrade recycling and waste-to-resource initiatives. *Example:* Composting units in the smart city turn organic wastes into fertilizer for agriculture.
5. **Green Building Design**

Circular principles encourage the use of modular, recyclable construction materials, while smart technology helps in achieving energy- and water-saving efficiency of buildings. *Example:* Smart house with energy-saving appliances plus solar panels combined with friendly recycling use of construction materials.

Application of circular economy principles in a smart city plan makes urban communities take the lead in fighting global pressures such as climate change, material scarcity, and increased population pressure.

12.3.2.4 Blockchain and IoT for Circular Supply Chains

Toward a circular economy entails the change in direction from traditional supply chains into a circular supply chain, that is, CSC. These are reusing resources with minimal waste, in such a way as to advance the sustainability agenda. Two significant technologies, namely, blockchain and IoT, support and hone such circular supply chains in that it addresses all the intricacies entailed by the resource management and materials and processes across its entire life cycle [13].

(i) Role of Blockchain in Engineering Circular Supply Chains

1. **End-to-End Traceability**
   Blockchain technology will ensure complete transparency along supply chains as each transaction and movement of materials is recorded on a secure, decentralized ledger. *Application:* Tracking the life cycle of recovered materials, in this case, recycled plastics and metals—from collection up to reintroduction into new products.
2. **Secure Data Sharing**
   This technology facilitates secure data interchange between various stakeholders like manufacturers, recyclers, and logistics providers so that there is transparency, and people gain trust on the blockchain technology. Its application will be relevant when it will certify the utilization of recycled materials or inputs sourced ethically in an engineering practice.
3. **Smart Contracts for Automation**
   Smart contracts automatically run such critical operations regarding the specific predefined rules when it is payment execution and related compliance check. *Application:* Payment to a recycler once the incoming materials qualify for reuse in further manufacturing processes.

4. Product Life Cycle Documentation
Blockchain-based digital passports are storing life cycle data, manufacturing details, maintenance records, and recycling pathways. *Application:* Assisting designers engineers to design products that have capabilities to easily disassemble or be repaired or for easier recycling.

(ii) Role of IoT in Engineering Circular Supply Chains

1. Real-time monitoring IoT devices, for example sensors and RFID tags, will offer real-time data regarding material flows, inventory, and environmental conditions. *Application:* Condition monitoring of returned products to evaluate the suitability of remanufacturing or recycling them.
2. Resource management IoT systems monitor resources, predict demand, and optimize inventory to minimize wastage. *Application:* Factory production is adjusted in intelligent factories based on the presence of materials and their patterns of consumption using IoT-based data.
3. Condition-based maintenance IoT allows predictive maintenance through a monitoring of equipment performance as it reduces downtime and increases life cycle of assets. *Application:* Maximizing the reuse of parts, preventing failure of industrial equipment.
4. Optimized reverse logistics IoT supports an efficient collection, transportation, and processing of used products and waste. *Application:* IoT-enabled tracking of electronics returned for recycling shall ensure that they reach the certified facilities.

(iii) Blockchain–IoT Synergy in Circular Supply Chains

The combination of blockchain and IoT addresses critical challenges in the implementation of circular supply chains:

1. Improved Transparency
IoT captures real-time information on materials and processes, and blockchain makes sure that such information is shared securely and traced. For instance, IoT sensors capture the quality of recyclable materials, and blockchain captures their origin and processing history.

2. Better Decision-Making
IoT devices offer actionable insights, while blockchain ensures integrity of data to enable better decision-making. *Example:* Engineers will be able to utilize blockchain-verified IoT data in the design of products that have higher recyclability.

3. Automation and Efficiency
IoT sensors activate smart contracts on the blockchain, thus automating processes such as waste collection or quality inspection. *Example:* IoT-enabled smart bins inform recycling facilities once they are full. Blockchain verifies the collection and sorting of materials.

4. Closed-Loop Systems
Blockchain and IoT allow closed-loop systems, allowing tracking of materials from raw material acquisition to end-of-life and return for reuse. *Example:* IoT tracks component usage in an industrial machine; blockchain records disassembly and reuse detail.

(iv) Engineering Applications in the Circular Economy

1. **Sustainable Manufacturing**

   IoT provides for the effective use of resources; on the other hand, blockchain verifies that inputs come from sources that are sustainably sourced or recycled. *Example:* Automotive industries apply IoT and blockchain in tracing recycled metals from scrap yards to production in new vehicles.

2. **Smart Energy Systems**

   IoT monitors consumption in real time, while blockchain supports decentralized trading of renewable energy. *Example:* Factories use IoT to optimize consumption of renewable energy, while blockchain tracks energy credits through circular operations. Eco-Design and Product Innovation Engineers use the data from blockchain to develop products with minimal environmental impact during their use, while IoT ensures monitoring. *Example:* IoT sensors monitor modular building components, and their reuse in new constructions is tracked using blockchain.

3. **Waste-to-Resource Innovations**

IoT discovers high-value waste streams and records their conversion into new materials on the blockchain. For example, urban food waste transformed into biogas IoT tracks the process and gets verified through blockchain.

The combination of blockchains and IoT enables disruptive forces in the making of circular supply chains, fostering improvement in the transparency, efficiency, and sustainability of managing sources. This combination allows industries to live up to the principles of the circular economy and develops robust systems that are free of waste, conserve resources, and foster innovation for a future world [14].

12.4 CONCLUSION

The shift toward a CE is an urgent imperative, as well as a challenging engineering task. Adoption of leading-edge technologies such as IoT, AI, blockchain, and sustainable manufacturing practices supports advancements, but significant barriers still exist. These are high costs of implementation, challenges of scalability, fragmented collaboration among different stakeholders, and the need for all-inclusive regulatory frameworks. Tackling these issues is going to require a multidisciplinary approach. Engineers are challenged to design the systems with modularity, high repairability, and resource efficiency on top of their list. Essential collaboration platforms that bring various stakeholders together, including government, industry, and other community organizations, must be forged. Breakthroughs in digital twins, machine learning, and even bio-inspired materials can spur a new wave of management of resources, reuse, and recyclability. Looking forward, engineering in the circular economy will be focused on developing scalable, intelligent systems to optimize resources, improving global interoperability to make cross-border material reuse easier, and using AI-driven predictive models to forecast material flows and minimize waste. By solving these problems, and with the help of the technological innovations, the circular economy should become a basic element of sustainable development, encouraging resilience, innovation, and environmental stewardship for coming generations.

REFERENCES

1. Lacy, P., Long, J., & Spindler, W. 2020. *Circular Economy Handbook: How to Build a More Resilient, Competitive, and Sustainable Business*. Palgrave Macmillan.
2. Stahel, W. R. 2019. *The Circular Economy: A User's Guide*. Routledge. https://doi.org/10.4324/9780429054010
3. Braungart, M., & McDonough, W. 2009. *Cradle to Cradle: Remaking the Way We Make Things*. North Point Press.
4. Kirchherr, J., Reike, D., & Hekkert, M. 2017. Conceptualizing the Circular Economy: An Analysis of 114 Definitions. *Resources, Conservation & Recycling*, 127, 221–232.
5. Geissdoerfer, M., et al. 2017. The Circular Economy—a New Sustainability Paradigm? *Journal of Cleaner Production*, 143, 757–768.
6. Korhonen, J., Nuur, C., Feldmann, A., & Birkie, S. E. 2018. Circular Economy as an Essentially Contested Concept. *Journal of Cleaner Production*, 175, 544–552. 3.
7. World Economic Forum. 2021. *The Circular Economy: A Transformative Path to Sustainability*. Report.
8. European Environment Agency (EEA). 2016. *Circular Economy in Europe: Developing the Knowledge Base*. EEA Report No. 2/2016.
9. McKinsey & Company. 2022. *The Next Frontier in Circular Economy: How Technology and Innovation Can Drive Sustainability*. New York: McKinsey & Company.
10. *The Ellen MacArthur Foundation's Circular Economy Case Studies: Comprehensive Examples of Circular Practices Across Industries*, www.ellenmacarthurfoundation.org.
11. Accenture Strategy. 2014. *Circular Advantage: Innovative Business Models and Technologies to Create Value in a World without Limits to Growth*. Accenture.
12. Preston, F. 2012. *A Global Redesign? Shaping the Circular Economy* (EERG BP 2012/02). Chatham House. https://www.chathamhouse.org/2012/03/global-redesign-shaping-circular-economy.
13. Rehman Khan, S. A., Yu, Z., Sarwat, S., Godil, D. I., Amin, S., & Shujaat, S. 2021. The Role of Block Chain Technology in Circular Economy Practices to Improve Organisational Performance. *International Journal of Logistics Research and Applications*, 25(4–5), 605–622. https://doi.org/10.1080/13675567.2021.1872512
14. Moreau, V., et al. 2017. Coming Full Circle: Why Social and Institutional Dimensions Matter for the Circular Economy. *Journal of Industrial Ecology*, 21, 497–506, https://doi.org/10.1111/jiec.12598

13 Synthesis and Future Directions on Circular Economy

Jeswin Jose, Nabanita Ghosh, Simran Hemanshu Naik, K. Kritin Nanjappa, and Maulika Premal Desai

13.1 INTRODUCTION

In an era of alarming environmental crises, climate change, and resource scarcity, the transformation from a linear to a circular economy represents one of the most significant shifts in global economic practices since the Industrial Revolution. This systemic transformation is aimed at redefining growth and sustainability. It starkly contrasts the traditional "take–make–dispose" approach. By following circular principles, companies can lessen their dependence on unstable raw material prices and supply chain interruptions by separating economic growth from resource consumption. Companies can improve customer satisfaction, obtain essential insights about consumer behavior, and possibly foster brand loyalty by transitioning to product leasing and service models.

Lack of capital is the major barrier in moving toward circular business models. The transformation process of circular models requires heavy investments (Mähönen 2019). So, there is a need for a stable cash flow. It is seen that many stakeholders are contributing to different roles in the circular business models (Haase, Bernegger, and Meslec 2024). It is reported that the circular economy enables the creation of shared value and strengthens the trust of investor and stakeholder relationships (Kretschmann 2022). In this context, it is vital to understand the factors which affect investor commitment (Mähönen 2019).

In a data-driven world, artificial intelligence applications and digital financial technologies play an important role in promoting sustainable investments (Grégoire and Guay 2023). There is a correlation between technology adoption and the success of circular business models (Kalogiannidis et al. 2022). However, this is also associated with several risks in terms of security and privacy (Morgan 2022). In data-driven operations, fuzzy logic has a vital role in the hands of corporates and marketers. It is reported that fuzzy logic can assist in better identifying new customers while considering the diversity of data and inaccuracy (Bashiri, Mirhosseini-Alizamini, and Nasrabadi 2020; Meier and Werro 2007). It helps provide customers with accurate information about preferences (Liu 2010). As the data is at the heart

DOI: 10.1201/9781003538837-13

of every operation, fuzzy logic is a critical element of business intelligence (Krishna and Supriya 2001).

As it is evident that fintech and digital technologies play a crucial role in circular investments, it is beneficial to understand the implications of digital financial literacy (DFL) is vital to ensure wise and independent decisions regarding sustainable investments. It is reported that a correlation exists between financial literacy and the adoption of ESG (environmental, social, and governance) investments (Misra, Bohra, and Sharma 2024). Also, it is informed that misconceptions are considered to be one of the reasons for a lower propensity to invest in socially responsible investments and can also be related to lower levels of financial literacy (Meunier and Ohadi, 2022). There is a strong positive influence on investor financial literacy with their sustainability information in making decisions on investments (Syahfi 2023). Financial literacy and environmental factors are key to determining one's sustainable investing (Gutsche, Wetzel, and Ziegler 2023).

In this context, the study explores the factors which can affect sustainable investment behavior in the circular economy context. Also, it discusses the importance of DFL in empowering investors to make wise and independent investment decisions in the circular economy.

13.2 OBJECTIVES OF THE STUDY

1. To examine the effect of awareness of ESG practices on sustainable investment behavior.
2. To study the effect of perception of sustainable investment on sustainable investment behavior.
3. To examine the effect of environmental considerations on sustainable investment behavior.
4. To examine the combined effect of awareness of ESG practices, perception of sustainable investment, and environmental considerations on sustainable investment behavior.
5. To discuss the implications of digital financial literacy in empowering investors.

13.3 LITERATURE REVIEW AND HYPOTHESIS DEVELOPMENT

13.3.1 Circular Business Models and Value Creation

The concept of circular economy is seen as the key to improving efficiency in production, logistics, and distribution by way of waste management, recycling, renovation technologies, and creating new business models (Apalkova et al. 2021). A lack of capital is the major barrier to a circular business model. The transformation of a linear to circular model requires immense financing in terms of distribution, inventory, production, logistics, and other costs. Also, it requires adjustments in product life cycles and resource management. The gap between the need for capital and the generation of cash flow is a major barrier in the circular economy. Considering this scenario, it is important to identify the factors that drive investor commitment in a circular economy (Mähönen 2019).

13.3.2 Fintech and Sustainable Finance

It is reported that fintech can promote sustainable finance, but it has risks (Morgan 2022). Investors are moving their capital into ESG (Environmental Social Governance)-oriented funds at an unprecedented pace; businesses are changing by embracing the technology and adapting to the changes. It is seen that fintech companies play a crucial role in developing the circular economy (Grégoire and Guay 2023). Digitalization can transform the economy into a circular economy. It is reported that there is a correlation between technology adoption and the success of circular economies. So, it is evident that digitalization is pivotal in the development of circular business models (Kalogiannidis et al. 2022).

13.3.3 Perception of Sustainable Investment

In the context of circular economy management, it is reported that different stakeholders are contributing in different roles in the circular economy. So, it is essential to trace these stakeholders and support them in contributing to a circular economy (Haase, Bernegger, and Meslec 2024). Word of mouth, perceived financial performance, risk performance, and financial literacy are the most important determinants of individuals' sustainable investments (Gutsche, Nakai, and Arimura 2021). It is shown that when the firm has higher levels of transparency and reputation, ESG performance strongly influences value creation (Ma, Pan, and Suardi 2023). So, in this context, exploring perception from the investor's side is vital.

13.3.4 Awareness of ESG Practices

Financial literacy is reported to be positively correlated with the adoption of ESG investments. It is important to acknowledge that financial education programs under ESG or CSR (Corporate social responsibility) strategy also played a crucial role in promoting financial literacy and financial inclusion (Waliszewski 2023). It is seen that a lack of proper awareness and financial literacy adversely affects socially responsible investments. It shows a strong positive influence on investor financial literacy with their sustainability information in making investment decisions (Syahfi 2023). Financial literacy and environmental factors are key factors in determining one individual's sustainable investment (Gutsche, Wetzel, and Ziegler 2023). There is another concept called sustainable finance literacy, which is conceptualized as the skill and knowledge in terms of understanding and evaluating financial products according to sustainability considerations. It is reported that those who show high levels of financial literacy may show low levels of sustainable finance literacy. Hence, it is necessary to note that awareness of sustainable finance is important in the ownership of sustainable products (Filippini, Leippold, and Wekhof 2024). So, it is important to consider the elements of sustainable finance with financial literacy.

13.3.5 Environmental Consideration

It is reported that sustainable value creation is central to companies and sustainability reporting. This integrated reporting gives a comprehensive view, which helps in internal decision-making. This information is meant as integrated information that helps create higher sustainable value. This information helps companies establish sustainable decision-making behaviors (Esch, Schnellbächer, and Wald 2019). It is evident that the COVID-19 pandemic has pushed investors to give more value to environmental and social responsibilities (Mavlutova et al. 2021). In this context, environmental considerations can strongly influence sustainable investing.

The following hypotheses are proposed in the light of the existing literature:

H1: Awareness of ESG practices has a significant effect on sustainable investment behavior.

H2: Perception of sustainable investment has a significant effect on sustainable investment behavior.

H3: Environmental considerations have a significant effect on sustainable investment behavior.

H4: Awareness of ESG practices, perception of sustainable investment, and environmental considerations have a significant combined effect on sustainable investment behavior.

13.4 RESEARCH METHOD

13.4.1 Sample

The target population of this study was existing and potential investors in the stock markets, who falls in the age group of 16 to above 50 years. The study was conducted in the Indian context through an online platform. Since the exact number of existing and potential investors in the stock markets, a convenient sampling technique was adopted.

13.4.2 Data Collection

The survey was conducted through online mode using Google forms in the Indian context. A Google form was circulated among the participants. The questionnaire was divided into sections based on criteria such as demographics, environmental considerations, perception of sustainable investment, awareness of ESG practices, and sustainable investment behavior. The analysis aimed to characterize the role of perception of sustainable investment, environmental consideration, and awareness of ESG practices related to sustainable investment behaviors of respondents than elaborate a proper evaluation scale for describing these factors related to investment behaviors (Walczak et al. 2021). Responses were gathered from 203 respondents consisting largely of working professionals and college students who invest in the stock market.

The description of the questionnaire items is presented in Table 13.1.

TABLE 13.1
Questionnaire Items

Awareness of ESG Practices	
1	How well do you understand the concept of a circular economy in the context of investment?
2	Do you check a company's disclosure on its ESG practices before investing in it?
3	Are you familiar with companies that are actively scaling up their recycling processes?
4	Have you observed any financial benefits for companies that have embraced circular practices?
Perception of Sustainable Investment	
1	How do you perceive the potential of sustainable investments in the near future?
2	How do you perceive the risk associated with investing in circular economy initiatives compared to traditional investments?
3	Do you take active interest in understanding a company's roadmap toward sustainability?
Environmental Consideration	
1	To what extent have environmental factors influenced your investment decisions in the past?
Sustainable Investment Behavior	
1	What is your portfolio's current allocation toward sustainable investments?
2	Are you planning to increase your allocation toward sustainable investments over the next five years?
3	Are you willing to invest in companies that are moving toward sustainability but are loss-making at present?

13.4.3 Methods of Data Collection

13.4.4 Conceptual Model

The model explores the relationship between the four variables—awareness of ESG practices, perception of sustainable investment, environmental considerations, and sustainable investment behavior. Awareness, perception, and environmental consideration are the independent variables. Sustainable investment behavior is the dependent variable. Figure 13.1 represents the conceptual model.

13.5 RESULTS AND DISCUSSION

13.5.1 Data Analysis

The data analysis was performed in two steps. In the first step, the demographic profile of the participants was tabulated based on their age, professional status, and educational qualification. In the second step, simple linear regression and multiple regression were conducted to test the hypothesis. The statistical analysis was conducted using the Jamovi 2.4.8 software for Windows (The Jamovi Project 2023).

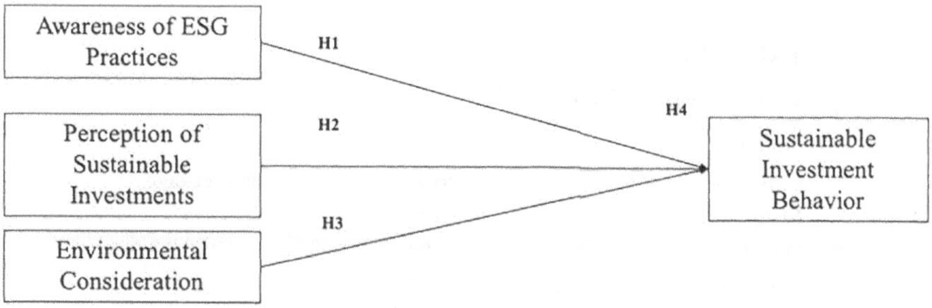

FIGURE 13.1 Conceptual model.

13.5.1.1 Demographic Profile

A total of 202 participants who are Indian citizens aged 15 to above 50 years responded to the survey. Of these, 50.5% were male, 48.5% were female, and 1% preferred not to say. Regarding age, 5.9% of participants are in the age group of 16–18, 87.1% of participants are in the 19–30 group, 3% in the 31–50 group, and 4% in the above 50 group. Regarding occupation, 88.6% were students, 8.9% were working professionals, 0.5% were engaged in business, and 2% were engaged in other activities. Regarding household size, 4.5% comes from 1–2, 68.3% from 3–4, 20.3% from 5–6, and 6.9% from more than 6 member households. With regard to marital status, 8.9% are married and 91.1% are unmarried.

The demographic profile of the participants is given in Table 13.2.

13.5.1.2 Hypothesis Testing

H1: Awareness of ESG practices has a significant effect on sustainable investment behavior.

The regression analysis shows a low R-square value of 0.0599, which indicates that there is only a small percentage of variation in sustainable investment behavior that may be linked to the awareness of ESG practices. The low R-squared value, however, may explain that even though there is a connection between the two variables, it is probable that extraneous variables exert a greater influence on the investor when it comes to making investment decisions. In terms of the model coefficients, the awareness (AWA) has a coefficient of 0.201. This coefficient provides a more nuanced understanding of the connection between the two variables. The awareness coefficient describes that for every unit increase in the awareness of ESG practices, there is a simultaneous increase of 0.201 units in investments toward sustainable options. In addition to this, the t-statistic of 3.57 and a p-value less than 0.001 indicate a statistically significant link between the two variables, hence implying the relevant statistical connection between being aware of ESG practices and choosing sustainable investments.

H2: Perception of sustainable investment has a significant effect on sustainable investment behavior.

TABLE 13.2
Demographic Profile

Gender		
Category	**Frequency**	**Percentage**
Male	102	50.5
Female	98	48.5
Prefer not to say	2	1
Age		
Category	**Frequency**	**Percentage**
16–18	12	5.9
19–30	176	87.1
31–50	6	3
51 and above	8	4
Occupation		
Category	**Frequency**	**Percentage**
Student	179	88.6
Working professional	18	8.9
Business	1	0.5
Other	4	2
Marital Status		
Category	**Frequency**	**Percentage**
Married	18	8.9
Unmarried	184	91.1
Household Size		
Category	**Frequency**	**Percentage**
1–2	9	4.5
3–4	138	68.3
5–6	41	20.3
More than 6	14	6.9

The linear regression which tested the relationship between investor perception of sustainable investments and investing in sustainable options showed a relatively low R-squared value of 0.107. This signifies that the variation between investor perception of sustainable investment and investing in sustainable options can only be minimally explained. The perception variable (PER) has a coefficient of 0.197. This implies that with every 0.197 units rise in the investment toward sustainable options, there is a unit increase in the positive perception. In addition, the relationship's statistical significance is given by the t-statistic of 4.91 and a p-value of less than 0.001. These values imply that the variables in question are statistically significant.

H3: Environmental considerations have a significant effect on sustainable investment behavior.

A linear regression analysis examining the relationship between environmental considerations and investing in sustainable options showed a low *R*-squared value of 0.0168. The model minimally explains the variation in the influence environmental consideration may have on sustainable investment decisions. The environmental consideration variable (ENV) has a coefficient of 0.231. This means that for every unit increase in the influence of environmental considerations, there is a simultaneous increase of 0.231 units in investment toward sustainable options. However, it is important to note that the *t*-statistic of 1.85 and *p*-value of 0.066 make the relationship between the variables less statistically significant. These values suggest the possibility that the observed connection between the two variables in question could be due to random chance.

H4: Awareness of ESG practices, perception of sustainable investment, and environmental considerations have a significant combined effect on sustainable investment behavior.

The relationship between three independent variables (perception of sustainable investment, environmental consideration, and awareness of ESG practices) and one dependent variable (sustainable investment behavior) was analyzed using multiple regression. *R*-squared value was 0.120. This model accounts for a comparatively small amount of the variation in sustainable investment behavior that can be attributed to the interaction between environmental considerations, how these investments are perceived, as well as awareness of ESG practices. Perception (PER) is 0.331, awareness (AWA) is −0.207, and the coefficient for environmental consideration (ENV) is 0.029. These coefficients show that there is a corresponding rise in investment toward sustainable choices of 0.029, 0.331, and −0.207 units for each unit increase in environmental considerations, positive perception of sustainable investments, and awareness of ESG practices, respectively. The statistically substantial correlation observed between perception and investment (*t*-statistic: 3.527, *p*-value <0.001) implies that their perceptions of these assets significantly influence investors' decisions to invest in sustainable investments. The *p*-value (0.099) indicates a marginally insignificant link despite the negative awareness coefficient, suggesting that further research is needed to understand the relationship between awareness and sustainable investment. In this case, the *t*-statistic (0.229) and *p*-value (0.819) show that, according to this model, there is no statistically significant evidence of a relationship between investment in sustainable projects and environmental considerations.

The study shows that perception, awareness, and environmental considerations explain 12% of the variance in investment behavior. In the context of social science research which is intended at checking the effects, it is reported that the *R*-squared value of at least 10% is acceptable when most of the predictors are significant (Ozili 2023). When individually, it is found that awareness and perception are statistically significant. Environmental considerations are not found to be statistically significant. It reiterates the concept of sustainable finance education and considers these factors.

Given this context, it is important to check the intersections between digital finance, sustainable investments, and building digital financial literacy. It is reported that digital finance is enhancing green innovations and, in turn, assisting in reducing financial constraints by way of alternatives. It is reported that digital finance can act as a booster for sustainability in terms of technological advancements (Chen et al. 2024). It is reported that green developmental finance is acting as a way to achieve SDG goals through financial inclusion and energy transition. It, in turn, promotes the transformation and upgradation of the industry (Xiao, Lin, and Wang 2024). It is reported that digital finance has a crucial role in framing a company's environmental responsibility practices (Yang et al. 2024). In the meantime, it is very important to note that digital technologies provide risk in terms of electronic waste, tax evasion, unauthorized cash flows, and asset exchanges regarding cryptocurrencies. Also, the concerns on financial security in terms of hacking and phishing. These risks need to be considered for the promotion of green digital finance. Here, digital financial literacy is also becoming crucial (Morgan 2022). There should be an intersection of finance, sustainability, and corporate responsibility in the digital era (Khalid, Irfan, and Srivastava 2024). On the contrary, as Industry 4.0 is characterized by AI, IoT, and automation, it makes a path for Industry 5.0 and stresses employee well-being and people and society. It calls for a win-win situation for both industry and society and, in turn, sustainable development. While designing such strategies, it is important to have collaborative efforts involving consumers, employees, shareholders, government, and society (Gigauri and Janjua 2023). Considering the unpredictable and imprecise nature of the finance domain, it is reported that fuzzy logic has a lot of scope (Deepan and Subbulakshmi 2023). However, there is a lack of usage of fuzzy logic. So, banks and other financial institutions regulators need to take this into consideration. It is reported that fuzzy logic can assist in financial planning regarding investment decision-making regarding stocks (Deepan and Subbulakshmi 2023). It is reported that fuzzy logic is highly beneficial in sustainability indices in making a global score to evaluate the sustainability of assets (Liern and Pérez-Gladish 2017).

13.6 CONCLUSION AND IMPLICATIONS

It is evident that while designing these circular business models, it is very important to understand investor behavior, because the transition of investors from a linear consumption pattern requires a behavioral change. It is also vital because such models require huge investments, and investor commitment becomes a determining factor. An investor-oriented approach is to be taken along with balancing the other stakeholders. Also, there should be a more effective usage of big data and artificial intelligence so as to gauge the consumers' and investors' behavior. From the investor perspective, it is very important to build digital financial literacy to ensure the independence of the decisions. Also, it is essential to build the skill as it is considered a new-age skill. At the same time, there needs to be a strong regulatory environment and uniformity among the regulations across the countries.

13.7 LIMITATIONS AND RECOMMENDATIONS

The study is an initial exploratory study to understand the role of perception, awareness, and environmental considerations in sustainable investments. Although the study is able to give some primary insights of investment behavior in a circular economy, the study is suffering from limitations. First, the study was exploratory in nature; it is not intended to capture the investment behavior fully, and it lacks a comprehensive approach. Also, the use of standardized scales can provide better insights regarding behavior. Also, the data collected was cross-sectional in nature, and it has limited representation. Future researchers can consider this an initial attempt to discuss the factors affecting investment behavior in circular economy and sustainable investments. The lower explained variance opens up ways for future researchers to explore the factors and explore more factors.

The world is striving toward sustainability. The circular economy is primarily viewed as an alternative model in place of a linear system in achieving sustainable development (Suchek, Ferreira, and Fernandes 2022). The circular economy is necessary to solve many environmental and social challenges (Nikolaou, Jones, and Stefanakis 2021). But operationalization is the key to the success of all the efforts to flourish. The circular economy involves organizations, customers, society, and the ecosystem (Alhawari et al. 2021). So, there needs an effort which is essential at both micro- and macrolevels. So, in the operationalization of the circular economy, it becomes very important (Liu, Zhu, and Feng 2019). So, in order to complement the circular economy efforts, there is a need for collaborative efforts. It is instrumental in bringing sustainability and achieving the SDGs.

REFERENCES

Alhawari, Omar, Usama Awan, M. Khurrum S. Bhutta, and M. Ali Ülkü. 2021. "Insights from Circular Economy Literature: A Review of Extant Definitions and Unravelling Paths to Future Research." *Sustainability* 13 (2): 859. https://doi.org/10.3390/su13020859.

Apalkova, Viktoriia, Sergiy Tsyganov, Tetiana Chernytska, Nataliia Meshko, and Nadiia Tsyganova. 2021. "Evaluating the Economic and Ecological Effects of Investment Projects: A New Model and its Application to Smartphone Manufacturing in Europe." *Investment Management and Financial Innovations/Investment Management & Financial Innovations* 18 (4): 252–265. https://doi.org/10.21511/imfi.18(4).2021.22.

Bashiri, Abbas, Seyed Mehdi Mirhosseini-Alizamini, and Mohammad Mehdi Nasrabadi. 2020. "Application of Fuzzy Logic for Advertising Marketing Campaigns." *Control and Optimization in Applied Mathematics (COAM)*, Payame Noor University: 25–37. https://doi.org/10.30473/COAM.2021.58392.1160.

Chen, Wei, Guzi Arn, Hongti Song, and Yu Xie. 2024. "The Influences of Digital Finance on Green Technological Innovation in China's Manufacturing Sector: The Threshold Effects of ESG Performance." *Journal of Cleaner Production* (June): 142953. https://doi.org/10.1016/j.jclepro.2024.142953.

Deepan, G. Surya, and T. Subbulakshmi. 2023. "Financial Planning and Decision Making for Students Using Fuzzy Logic." (September): 159–171. https://doi.org/10.1002/9781394175130.ch10.

Esch, Martin, Benedikt Schnellbächer, and Andreas Wald. 2019. "Does Integrated Reporting Information Influence Internal Decision Making? An Experimental Study of Investment Behavior." *Business Strategy and the Environment* 28 (4): 599–610. https://doi.org/10.1002/bse.2267.

Filippini, Massimo, Markus Leippold, and Tobias Wekhof. 2024. "Sustainable Finance Literacy and the Determinants of Sustainable Investing." *Journal of Banking & Finance* 163 (June): 107167. https://doi.org/10.1016/j.jbankfin.2024.107167.

Gigauri, Iza, and Laeeq Razzak Janjua. 2023. "Digital and Sustainable Products to Achieve Sustainable Business Goals Along the Path to Industry 5.0." Emerald Publishing Limited eBooks, 25–40. https://doi.org/10.1108/978-1-83753-190-520231003.

Grégoire, Vincent, and Kevin Guay. 2023. "Circular Economy: A Fintech Driven Solution for Sustainable Practices." In *Fintech and Sustainability*, 149–168. https://doi.org/10.1007/978-3-031-40647-8_8.

Gutsche, Gunnar, Miwa Nakai, and Toshi H. Arimura. 2021. "Revisiting the Determinants of Individual Sustainable Investment—the Case of Japan." *Journal of Behavioural and Experimental Finance* 30 (June): 100497. https://doi.org/10.1016/j.jbef.2021.100497.

Gutsche, Gunnar, Heike Wetzel, and Andreas Ziegler. 2023. "Determinants of Individual Sustainable Investment Behavior—a Framed Field Experiment." *Journal of Economic Behavior & Organization* 209 (May): 491–508. https://doi.org/10.1016/j.jebo.2023.03.016.

Haase, Matthias, Heinz Bernegger, and Mihaela Meslec. 2024. "Stakeholder Mapping for Value Creation of New Circular Business Models." In *Lecture Notes in Civil Engineering*, 325–334. https://doi.org/10.1007/978-3-031-57800-7_30.

The Jamovi Project. 2023. "jamovi". (Version 2.4) [Computer Software]. https://www.jamovi.org.

Kalogiannidis, Stavros, Dimitrios Kalfas, Fotios Chatzitheodoridis, and Stamatis Kontsas. 2022. "The Impact of Digitalization in Supporting the Performance of Circular Economy: A Case Study of Greece." *Journal of Risk and Financial Management* 15 (8): 349. https://doi.org/10.3390/jrfm15080349.

Khalid, Fahad, Muhammad Irfan, and Mohit Srivastava. 2024. "The Impact of Digital Inclusive Finance on ESG Disputes: Evidence from Chinese Non-Financial Listed Companies." *Technological Forecasting & Social Change/Technological Forecasting and Social Change* 204 (July): 123415. https://doi.org/10.1016/j.techfore.2024.123415.

Kretschmann, J., S. Krause, and O. Drusche. 2022. "Circular Business Model Innovation In Mining: Towards a Sector-Specific Framework—SME Annual Meeting." *OneMine*. https://onetunnel.org/documents/circular-business-model-innovation-in-mining-towards-a-sector-specific-framework-sme-annual-meeting.

Krishna, P. Radha, and Kumar De Supriya. 2001. "A Fuzzy Approach to Build an Intelligent Data Warehouse." IOS Press. https://ip.ios.semcs.net/articles/journal-of-intelligent-and-fuzzy-systems/ifs00145.

Liern, Vicente, and Blanca Pérez-Gladish. 2017. "Companies' Selection Methods for Inclusion in Sustainable Indices: A Fuzzy Approach." In *Studies in Systems, Decision and Control*, 365–380. https://doi.org/10.1007/978-3-319-69989-9_21.

Liu, Chin-Hung. 2010. "A Group Decision-Making Method with Fuzzy Set Theory and Genetic Algorithms in Quality Function Deployment." *Quality and Quantity* 44 (6): 1175–1189. https://doi.org/10.1007/s11135-009-9304-1.

Liu, Junjun, Qinghua Zhu, and Yunting Feng. 2019. "The Circular Economy and Green Supply Chain Management." Edward Elgar Publishing eBooks. https://doi.org/10.4337/9781786434272.00007.

Ma, Ruichen, Xiaofei Pan, and Sandy Suardi. 2023. "Shareholder Value Maximization via Corporate ESG Performance: Evidence from Mergers and Acquisitions in China." *Applied Economics* (December): 1–17. https://doi.org/10.1080/00036846.2023.2291095.

Mähönen, Jukka. 2019. "Financing Sustainable Market Actors in the Circular Economy." Edward Elgar Publishing eBooks. https://doi.org/10.4337/9781788119283.00016.

Mavlutova, Inese, Andris Fomins, Aivars Spilbergs, Dzintra Atstaja, and Janis Brizga. 2021. "Opportunities to Increase Financial Well-Being by Investing in Environmental, Social and Governance with Respect to Improving Financial Literacy Under COVID-19: The Case of Latvia." *Sustainability* 14 (1): 339. https://doi.org/10.3390/su14010339.

Meier, Andreas, and Nicolas Werro. 2007. *A Fuzzy Classification Model for Online Customers. Informatica*. Vol. 31. Slovenian Society Informatica.

Meunier, Luc, and Sima Ohadi. 2022. "Misconceptions About Socially Responsible Investments." *Journal of Cleaner Production* 373 (November): 133868. https://doi.org/10.1016/j.jclepro.2022.133868.

Misra, Abhishek, N. S. Bohra, and Manu Sharma. 2024. "Impact of Financial Literacy Towards ESG Investing Among Salaried Employees." In *Advances in Logistics, Operations, and Management Science Book Series*, 274–284. https://doi.org/10.4018/979-8-3693-2849-1.ch018.

Morgan, Peter J. 2022. "Assessing the Risks Associated with Green Digital Finance and Policies for Coping with Them." In *Economics, Law, and Institutions in Asia Pacific*, 51–68. https://doi.org/10.1007/978-981-19-2662-4_3.

Nikolaou, Ioannis E., Nikoleta Jones, and Alexandros Stefanakis. 2021. "Circular Economy and Sustainability: The Past, the Present and the Future Directions." *Circular Economy and Sustainability* 1 (1): 1–20. https://doi.org/10.1007/s43615-021-00030-3.

Ozili, Peterson K. 2023. "The Acceptable R-Square in Empirical Modelling for Social Science Research." In *Advances in Knowledge Acquisition, Transfer, and Management Book Series/Advances in Knowledge Acquisition, Transfer and Management Book Series*, 134–143. https://doi.org/10.4018/978-1-6684-6859-3.ch009.

Suchek, Nathalia, João J. Ferreira, and Paula O. Fernandes. 2022. "A Review of Entrepreneurship and Circular Economy Research: State of the Art and Future Directions." *Business Strategy and the Environment* 31 (5): 2256–2283. https://doi.org/10.1002/bse.3020.

Syahfi, Salsabila Dwinda. 2023. *Why Do Some Young Investors in Indonesia Integrate Sustainable Investment in Their Portfolio? Review of Integrative Business and Economics Research*. Vol. 12. GMP Press and Printing.

Walczak, Damian, Leszek Dziawgo, Danuta Dziawgo, Michał Buszko, Jarosław Pawłowski, Agnieszka Żołądkiewicz-Kuzioła, and Dorota Krupa. 2021. "Attitudes and Behaviors Regarding Environmental Protection in the Financial Decisions of Individual Consumers." *Energies* 14 (7): 1934. https://doi.org/10.3390/en14071934.

Waliszewski, Krzysztof. 2023. "Socio-Economic Determinants of Financial Inclusion in Central and Eastern Europe during the COVID-19 Pandemic." *Przegląd Wschodnioeuropejski* 14 (1): 25–41. https://doi.org/10.31648/pw.9019.

Xiao, Yubo, Muxi Lin, and Lu Wang. 2024. "Impact of Green Digital Finance on Sustainable Development: Evidence from China's Pilot Zones." *Financial Innovation* 10 (1). https://doi.org/10.1186/s40854-023-00552-9.

Yang, Ping, Yanqin Lv, Xiaodan Chen, and Juan Lv. 2024. "Digital Finance, Natural Resource Constraints and Firms' Low-Carbon Behavior: Evidence from Listed Companies." *Resources Policy* 89 (February): 104637. https://doi.org/10.1016/j.resourpol.2024.104637.

14 Analyzing the Influence of Product Quality on Consumer Perception

A Comparative Analysis Between Rural Market of India and Developed Country Market

S. Parul, K. Taranjeet, K. Lohith, M. Venus, A. Christo, and Y. Mukhammadmurod

14.1 INTRODUCTION

Perception is considered as one of the most important factors of psychology which affects the decision-making behavior of an individual who decodes the various messages received from different platforms of the environment (Kotler & Armstrong, 2011). Consumer perception is crucial in determining whether a product is successful or unsuccessful, so it a hot topic of concern for the marketers to understand the nature or mood of potential customers/consumers as they form perceptions based on a variety of factors, including their previous experiences with the product or brand, the opinions of others, and the product's features and benefits (Inabo, 2025).

As a result, companies often invest significant resources in shaping consumer perceptions through advertising, branding, and other marketing efforts. It can be witnessed that during some previous decades, origin of brands (Indian or foreign) becomes a matter of concern for the consumer and getting significant attention from marketers, researchers and practitioners globally. Basically, consumers always show interest in knowing whether the product is produced by domestically possessed company or foreign-possessed company (Papadopoulos & Heslop, 1993). We are the witness of continuously changing business environment in the era of globalization with the blasting Asian economies such as India, China, and others (Wilson & Purushotaman, 2003). All these changes pressurize the distributors to know and understand the nature of different types of new markets across the boundaries (Golub & Hopkins, 2003). The revenue from the market for luxury goods in 2025

DOI: 10.1201/9781003538837-14

is estimated to be 387 billion dollars (Statista, 2022). Indian consumers have valued luxury products greatly as it portrays a status sign, with the luxury segment growing at the rate of 40%. Luxury brands have been accorded qualities such as craftsmanship, rarity, quality, uniqueness, to name but a few (Kautish et al., 2020).

Liberalization and free-trading, result of new economic policy which came into existence in 1991, give boost to the retail sector as global marketers compete in the market and provide *n* number of choices to the consumers and changed their living standards (Sabnavis, 2003). India is emerging as a most preferred destination for investments. Growth of foreign direct investment has increased 20 times in the last 20 years (Ministry of Commerce & Industry, FY 21–22). Inflow of FDI increases by 23% (USD 171.84 billion) as compared to FDI inflow in pre-COVID period (USD 141.10 billion) in India (*Economic Times*, 2022). At the same time, the retail scenario in India reflects several opportunities for local and global retailers as consumers' purchasing power and willingness to spend are on rise (Bhattacharya, 2007). Being among the world's economies with the quickest rate of growth, India is providing a great and strong platform to customers, distributors and manufacturers globally.

14.2 REVIEW OF LITERATURE

Consumer perception toward Indian and foreign products in the Indian retail industry has been a topic of interest in marketing and consumer behavior research. Various studies have examined the factors influencing consumer perceptions and preferences, as well as the impact of these perceptions on purchase decisions. Singh and Gupta (2017), in their study "Consumer Perception of Indian versus Foreign Products in the Indian Market: According to "A Comparative Study," consumer views are significantly affected by perceived quality, company reputation, and place of origin. According to the results, customers believe that international goods are more dependable and are of higher quality than Indian goods. Gupta and Sharma (2018) in their study, "Factors Influencing Consumer Perception toward Indian and Foreign Brands: A Comparative Study," found that perceived quality, brand image, and country of origin significantly affect consumer perceptions. They also found that foreign brands were perceived to have higher quality and better brand image compared to Indian brands. Priyadarshini and Jain (2019) in their study, "Consumer Perception and Buying Behaviour toward Foreign Brands: A Study of Indian Consumers," explored the factors and revealed that perceived quality, brand reputation, and social influence are important determinants of consumer perception. It also suggests that consumers perceive foreign brands as superior in terms of quality and reliability compared to Indian brands.[8]

Sharma and Sharma (2019) in their study, "Consumer Perception towards Indian and Foreign Brands: A Study of Indian Retail Sector," suggest that perceived quality, brand image, and product origin significantly affect consumer perceptions. The study also indicates that foreign brands are perceived as more fashionable and trendier compared to Indian brands. Singh and Verma (2020) in their study, "Consumer Perception towards Indian and Foreign Brands in the Indian Retail Market," investigate that perceived quality, brand image, and product origin significantly impact consumer perception. According to the report, people believe that international companies are more inventive and provide better value. Overall, the literature on consumer perception toward Indian and foreign products in the Indian retail industry consistently highlights the significance of factors such as perceived quality, brand image, and country of origin. Customers typically believe that foreign brands are better than Indian products in terms of quality, dependability, and innovation. These findings can help marketers and retailers understand consumer preferences and tailor their strategies to effectively position and promote their products in the Indian market.

14.3 OBJECTIVES OF THE STUDY

1. To gain insights about customer's perception about foreign and Indian brands.
2. To research the elements that affect consumers' purchasing decisions in the retail sector.
3. To find out the correlation between perception and factors affecting consumers buying behavior in retail industry.

14.4 RESEARCH METHODOLOGY

The research is based on descriptive type of study. It is related to the perception of consumers toward foreign and Indian retailing. One hundred consumers were surveyed using questionnaire. The data collected for the research study is based on primary data. Questionnaires were circulated to various consumers of Ludhiana. Convenience sampling is being used in this particular study. The responses of 100 consumers have been taken in this study.

14.5 LIMITATIONS OF THE STUDY

1. This study is limited to only 100 consumers of Ludhiana.
2. The study is not focusing on any specific segment of consumers as convenience sampling is being used.

14.6 DATA ANALYSIS AND INTERPRETATION

The respondents' demographic profile is shown in Table 14.1. Male respondents make up 43.3% of the sample, while female respondents make up 56.7%. Respondents between the ages of 18 and 24 make up the majority (63.9), followed by those between the ages of 25 and 34 (25.8%). The aforementioned data also makes it evident that students make up the bulk of responders (52.6%), followed by full-time employees. It can be seen from the data that 41.2% respondents have done their Master's degree and 37.1% have done their Bachelor's degree.

In the context of the retail industry, how does your belief in a difference in quality between Indian and foreign brands influence the importance you place on the perception of a brand's origin (Indian or foreign) when making purchasing decisions? (Table 14.2).

TABLE 14.1
Demographic Profiles

S.No.	Variable	Option	No. of Respondents	Percentage
1	Gender	Male	45	43.3
		Female	55	56.7
		Prefer not to say	0	0
2	Age	18–24	65	63.9
		25–34	25	25.8
		35–44	5	5.2
		45–54	3	3.1
		55 and above	2	2.1
3	Occupation	Student	51	52.6
		Employed (full-time)	24	21.6
		Self-employed	15	15.5
		Unemployed	9	9.3
		Housewife	1	1
4	Educational Qualification	High School or equivalent	16	16.5
		Bachelor's degree	36	37.1
		Master's degree	43	41.2
		Doctorate or higher	2	2.1

TABLE 14.2
Relationship between Factor Influencing and Perception of Brand's Origin

Consumer Perception	Factors Influencing Consumer Buying Behavior	*R*1	*R*2	*R*1–*R*2 = *D*	D^2
32	21	4	3	1	1
41	24	5	4	1	1
10	31	3	5	–2	4
8	20	1	2	–1	1
9	4	2	1	1	1
					$\Sigma D^2 = 8$

$$r = 1 - \frac{6\sum d2}{n(n2-1)} \longleftrightarrow r = 1 - \frac{6*8}{5(25-1)} \longleftrightarrow r = 1 - \frac{48}{5(24)}$$

$$\longleftrightarrow r = 1 - \frac{48}{120}$$

$$r = 1 - \frac{48}{120} \qquad r = 1 - 0.4$$

$$r = 0.6$$

TABLE 14.3
What factors influence your choice when buying products in the retail industry?

Variable	Scale	Rank
Price	Least influential 1	30
	Slightly influential 2	14
	Neutral 3	7
	Very influential 4	12
	Most influential 5	9
Quality	Least influential 1	12
	Slightly influential 2	26
	Neutral 3	5
	Very influential 4	16
	Most influential 5	13
Brand reputation	Least influential 1	8
	Slightly influential 2	13
	Neutral 3	32
	Very influential 4	6
	Most influential 5	13
Product features and specifications	Least influential 1	5
	Slightly influential 2	13
	Neutral 3	19
	Very influential 4	31
	Most influential 5	4
Convenience (e.g., store location, accessibility)	Least influential 1	17
	Slightly influential 2	6
	Neutral 3	9
	Very influential 4	7
	Most influential 5	33

From Table 14.2, it can be interpreted with the use of Spearman's Rank Correlation formula, the value of $r = 0.6$ which is less than 1, it means there is a positive correlation between consumer perception of brand's origin (Indian or foreign) and factors influencing consumer buying behavior in Indian retail industry. It indicates that consumers have made their mind for purchasing get influenced by their perception regarding brand's origin (Indian or foreign) which is affected by difference in quality between Indian and foreign brands in Indian retail industry.

Table 14.3 clearly shows that convenience (e.g., store location, accessibility) is the most influencing factors while making buying products in the retail industry. And product features and specifications is the second most influencing factors which affect the buying behavior of consumers which is followed by quality of the product. Price and brand reputation have least impact over the buying behavior of consumers while making purchasing decisions in Indian retail industry.

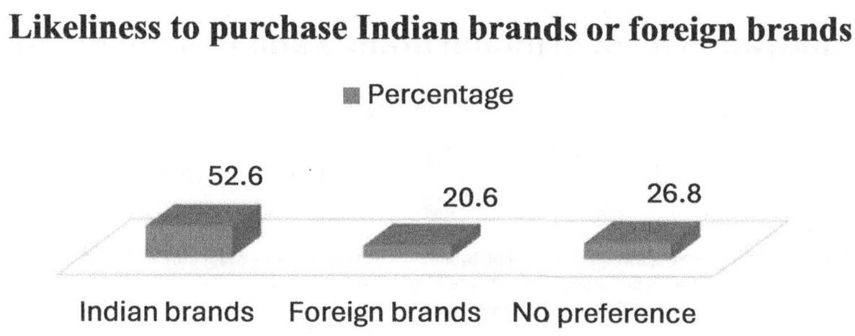

FIGURE 14.1 A bar diagram showing likeliness to purchase Indian brands or foreign brands.

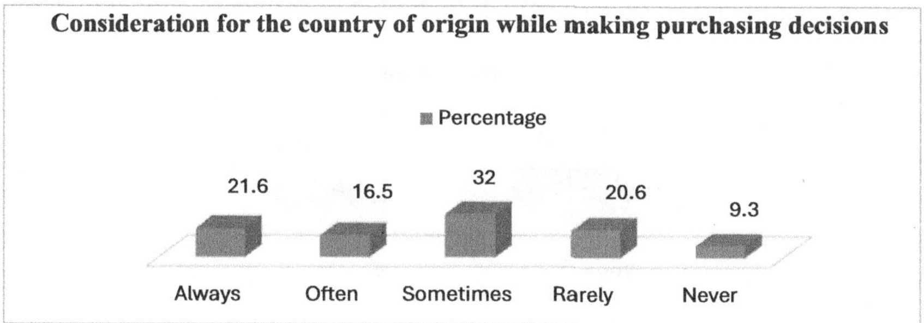

FIGURE 14.2 A bar diagram showing consideration for the country of origin while making purchasing decisions.

It can be easily analyzed from Figure 14.1 that 52% respondents prefer to purchase Indian brands over foreign brands, that is, 21% in the Indian retail industry. 27% respondents have no preference while making their purchasing decisions in Indian retail industry.

Figure 14.2 shows that a 32% respondent says that sometimes they consider the country of origin while making purchasing decisions. Around 22% are in favor of always check, 21% respondents are in favor of rarely they check the country of origin. It is followed by 16% those respondents who says often they check the country of origin and 9% says they never bothered about the country of origin while making purchasing decisions in the retail industry.

Figure 14.3 clearly depicts that majority of respondent (45%) says that foreign brands are more innovative in comparison to Indian brands (25%) in the retail industry; 18% respondents says there is no significant difference between Indian and foreign brands while considering the innovation level and 12% says that they are not sure about it.

From Figure 14.4, it can be analyzed that a majority of respondents (62%) say that they have changed their purchasing decisions based upon their perception of a brand origin (Indian or foreign) and 38% respondents say that they have not changed their purchasing decision based on their perception of a brand's origin (Indian or foreign).

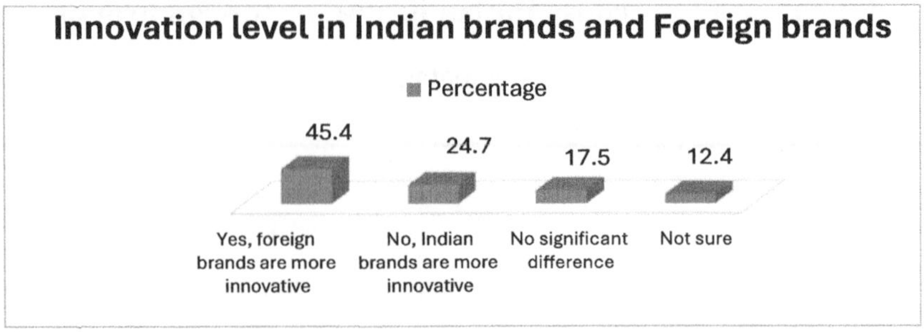

FIGURE 14.3 A bar diagram showing Innovation level in Indian brands & Foreign brands.

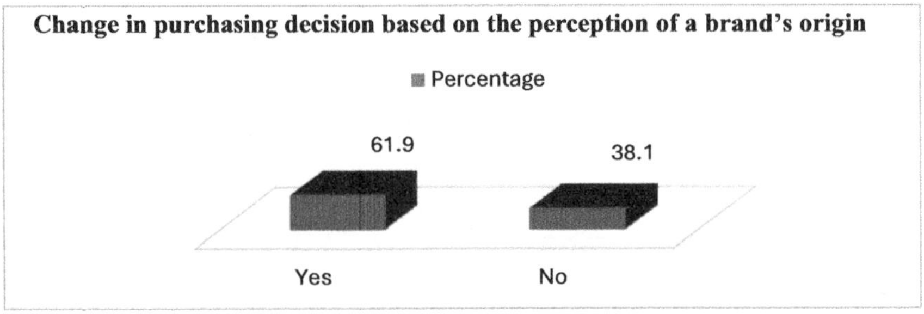

FIGURE 14.4 A bar diagram showing change in purchasing decision based on the perception of a brand's origin.

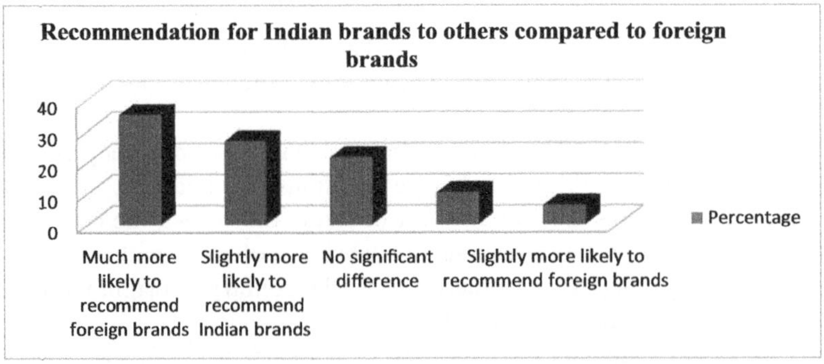

FIGURE 14.5 A bar diagram showing recommendation for Indian brands to others compared to foreign brands.

Figure 14.5 depicts that around 35% respondents are much more likely to recommend Indian brands to others as compared to foreign brands and 27% respondents are slightly more likely to recommend Indian brands to others as compared to foreign brands whereas 22% respondent say that there is no significant difference between Indian and foreign brands and only 6% say that they are in favor of highly recommendation of

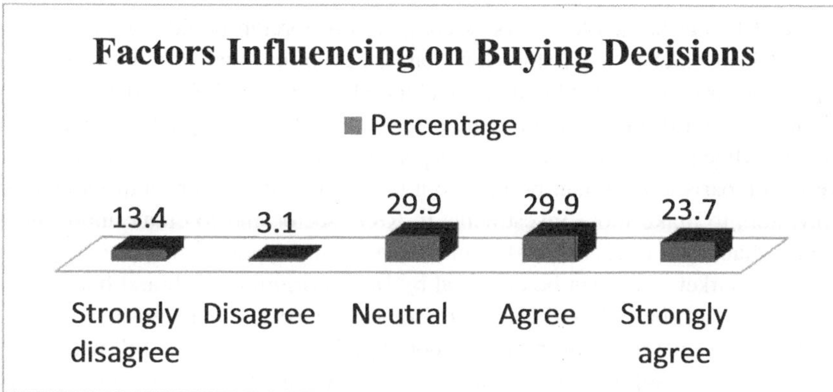

FIGURE 14.6 A bar diagram showing factors influencing on Buying Decisions.

foreign brands to other as compare to Indian brands and 10% respondents say that they slightly more prefer to recommend foreign brand over Indian brands.

Figure 14.6 presents the interpretation that around 30% respondents agree and 24% respondents strongly agree with the statement that "Factors like price, quality, and convenience have a stronger influence on my buying decisions than the perception of a brand's origin" and 13% respondents strongly disagree with the statement followed by 3% are those they also disagree with the statement. Whereas a 30% respondent say that they are neutral with the statement that "Factors like price, quality, and convenience have a stronger influence on my buying decisions than the perception of a brand's origin."

14.7 RESULT AND DISCUSSION

Major finding of the study is that consumer while making purchasing decisions is positively influenced by brand's origin (Indian or foreign). Convenience (e.g., store location, accessibility) and Product features and specifications are the most influencing factors while making buying decisions in the retail industry. Brand reputation and price are the least influencing factors for consumers while making buying decisions in the retail industry. Indian brands are the most preferred brands for most of the respondents in Indian retail industry. Majority of respondents believe that foreign brands are more innovative as compared to Indian brands in the retail industry. As per the analysis, respondents recommend Indian brands to others as compared to foreign brands in the retail industry. Majority of respondents consider price, quality, and convenience as factors having a stronger influence on buying decisions than the perception of a brand's origin. Quality is an important factor which influence consumers while making buying decisions in the retail industry.

14.8 CONCLUSION AND SUGGESTIONS

In conclusion, it has been identified that consumer's perception is positively influenced by brand's origin (Indian or foreign) and quality perspective and they prefer to recommend Indian brands to others as compared to foreign brands. It also has been found that Convenience (e.g., store location, accessibility) and Product features and specifications are the most influencing factors while making buying decisions in the retail industry and Brand reputation and price comes at last being an influencer to the consumers while purchasing decisions. As per the study, foreign brands are more innovative in comparison to Indian brands, so it is a suggestion to our retail industrialists and investors to make more investments in R&D sector and to create more innovations in Indian origin brands so that more marketing opportunities can be grabbed and greater market share can be acquired by Indian origin retail brand houses in the world total market share. It will also increase our exports and put a positive impact on Indian balance of payment. So being a labor-abundant and resource-full nation, India can use its skilled manpower to introduce more creativeness and innovation in the production sector in retail industry so that expenditure of Indian citizens will be utilized in strengthening the Indian economy only, not in the foreign pockets.

REFERENCES

Bhattacharya, S. "Retail Sector in India: Opportunities and Challenges." *Indian Journal of Marketing* 37, no. 2 (2007): 21–27.

Economic Times. "India's FDI Inflow Increased by 23% to USD 171.84 Billion in 2021–22," 2022. https://economictimes.indiatimes.com.

Golub, S., and Hopkins, M. (2003). "Understanding New Markets across Boundaries." *International Journal of Market Research* 45, no. 2 (2003): 123–135.

Gupta, P., and Sharma, R. "Factors Influencing Consumer Perception toward Indian and Foreign Brands: A Comparative Study." *Journal of Marketing Research* 55, no. 2 (2018): 133–148.

Inabo, S. *Customer perception: Definition, importance + how to improve it.* Zendesk, 28 February 2025. https://www.zendesk.com/in/blog/simple-guide-customer-perception/:contentReference[oaicite:4]{index=4}.

Kautish, S., Singh, S., and Kautish, P. "Qualities of Luxury Brands: Craftsmanship, Rarity, and Uniqueness." *Journal of Luxury Brand Management* 21, no. 3 (2020): 245–259.

Kotler, P., and Armstrong, G. *Principles of marketing* (14th ed.). Pearson Education, 2011.

Papadopoulos, N., and Heslop, L. A. (Eds.). *Product-country images: Impact and role in international marketing*. International Business Press, 1993.

Priyadarshini, S., and Jain, R. "Consumer Perception and Buying Behaviour toward Foreign Brands: A Study of Indian Consumers." *Asian Journal of Business and Management* 7, no. 4 (2019): 23–35.

Sabnavis, C. *India's economic reforms: The impact on retail sector*. Mumbai: KBI Publishers, 2003.

Sharma, M., and Sharma, S. "Consumer Perception towards Indian and Foreign Brands: A Study of Indian Retail Sector." *Retail Management Journal* 8, 1 (2019): 50–66.

Singh, R., and Gupta, A. "Consumer Perception of Indian versus Foreign Products in the Indian Market: A Comparative Study." *International Journal of Business and Management* 12, no. 3 (2017): 45–58.

Singh, P., and Verma, N. "Consumer Perception towards Indian and Foreign Brands in the Indian Retail Market." *International Journal of Retail & Distribution Management* 48, no. 5 (2020): 520–534.

Statista. "Luxury Goods Market Revenue Forecast Worldwide 2025," 2022. https://www.statista.com/statistics/forecast-luxury-goods-market-revenue-2025/.

Wilson, D., and Purushotaman, A. "The Impact of Globalization on Asian Markets." *Journal of International Business Studies* 34, no. 1 (2003): 1–16.

15 Circular Economy and Triple Bottom Line

A Case for India

Rahul Dhiman, Basit Ali Bhat, and Mohammad Shahparan

15.1 INTRODUCTION

The circular economy concept has caught the attention of the world as a way to provide a solution to resource scarcity as well environmental degradation (Korhonen et al., 2018). Its purpose is to alter production as well as consumption processes through increased resource efficiency along with a growth in the economy (Geissdoerfer et al., 2017). There are challenges in adapting circular economy, particularly in a developing country like India with complex socio economic conditions. There is a need to establish a stronger bond between the circular economy and sustainability for effective resource management (Geissdoerfer et al., 2017). The needs and expectations of the circular economy will support the reduction of product costs and increase the investment attractiveness of enterprises by reducing waste management costs (Korhonen et al., 2018). It will also help to create new jobs while simultaneously reducing the output of materials and energy waste. For developing countries such as India, it is imperative to bring together the sustainable development goals and the circular economy for balanced development (Dantas et al., 2021).

India faces several challenges in the areas of waste management, resource shortage, and environmental degradation that impede the circular economy adoption. The approach known as the circular economy has a strong correlation with the triple bottom line approach which takes into account the social, economic, and environmental implications (Aguiñaga et al., 2018). Economic, environmental, and social benefits can be realized through circular economy approaches (Geissdoerfer et al., 2017). It contributes to economic advancement, optimization of resource use, reduction of waste and pollution, and strengthening community health (Blomsma & Brennan, 2017). The integration of ideas of circular economy into corporate strategy and policies is imperative for achieving sustainable advancement. This research provides an analysis of how circular economic practices can be adopted within the Indian context bringing out the interactions with the triple bottom line and addresses the relevant policy gaps and institutional structures that need to be in place so as to promote circular economy practices which further enhances the chances of achieving sustainable development goals of the country.

DOI: 10.1201/9781003538837-15

15.2 THE CIRCULAR ECONOMY IN INDIA: A LANDSCAPE ANALYSIS

The circular economy has an outstanding potential to be included in India's development agenda by enhancing the life cycles of products and dealing with economic, environmental, and social issues. There literally is a consensus on the need to apply the principles of circular economy to achieve sustainability, reduce waste, and increase resource efficiency (De Angelis, 2018). However, these advantages are sometimes offset by the poor integration of the circular economy in sustainable development, partially warranting an emphasis on the oft-tokenistic social elements of CE-centered approaches (Kirchherr et al., 2017). There is no doubt that the circular economy and sustainability relate to each other in so many ways; however, there are fields, including aims, hierarchy of the issues, origins, and perception of them that differ (Geissdoerfer et al., 2017).

15.3 INTEGRATING TRIPLE BOTTOM LINE INTO CIRCULAR ECONOMY

In the recent past, there has been a teeming interest specifically in the concepts of circular economy and triple bottom line globally (Ferro et al., 2019). The driving force of triple bottom line is that it calls for businesses and institutions to not only account for their economic health but also consider their environmental and social performance (Miller, 2020). At the same time, the circular economy provides an integrative view in addressing pertinent contemporary challenges while fostering sustainable development of developing countries like India which is attempting to balance economic development, environmental protection, and social advancement (Gurin et al., 2015). The combination of triple bottom line and circular economy has moved up the agenda for India since it paves the way for a more sustainable development (Ahmed et al., 2022).

15.4 CHALLENGES AND OPPORTUNITIES OF TRANSITION TOWARD CIRCULAR ECONOMY

It is true that the circular economy concept has been embraced by many societies for a while; however, lack of clarity regarding what it is and what it encompasses is still a common phenomenon (Blomsma & Brennan, 2017). In developing economies like India, the absence of adequate supporting infrastructure acts as a bottleneck to its effective realization. Also, India seems to be focusing on the circular economics initiatives from a social and environmental perspective rather than from a commercial one which is paradoxical since it is a profit-oriented industry. Because of insouciance and lack of comprehension about the concepts of reuse and recycling in given industries, it gets difficult to implement policies aimed at development of circular economy which in most cases increases costs of going green (Hannon et al., 2016). This chapter investigates the challenges of the circular economy based upon extensive data and information gathered through analysis. The difficulties are further eminent into social, economic, and environmental dimensions, thus allowing for further enlightenment on the challenges through discussion.

15.5 SOCIAL CHALLENGES

The informal unorganized waste management is a critical issue in India (Kala et al., 2022). Consumer behavior of replacing products instead of repairing and recycling creates hurdles for circular economy success. Rural areas lack knowledge and facilities for waste management. Only 70% of the generated waste is collected and a smaller portion is treated (MSW Annual Report, 2020–21). Transitioning to circular economy may lead to job losses in automated industries due to technology investments. Recycled products are often considered inferior (Mwanza, 2021), and the lack of skill development programs for workers hinders circular economy implementation in India.

15.6 ENVIRONMENTAL CHALLENGES

In India, a large amount of plastic waste remains untreated, causing harm to the environment and hindering circular economy success. Only 30% of the 3.4 million tons of plastic waste produced annually is recycled (The Economic Times, 2023). Water scarcity is a pressing issue, with India projected to be the most affected country, affecting 1.7–2.4 billion people by 2050 (Outlook India, 2023). Increasing air pollution from vehicles, factories, and waste burning poses a significant challenge for circular economy goals. India ranks 4th among the world's most polluted cities, affecting air quality and ecosystems (The Indian Express, 2023). Implementing circular economy practices can reduce waste, deforestation, and ecosystem degradation, contributing to a sustainable future.

15.7 ECONOMIC CHALLENGES

The limited number of compulsory rules and regulations by the government hinders the smooth implementation of circular economy in India (Ghisellini et al., 2016). Due to poor policies and the absence of precise standards for evaluating circular economy success, government is less interested to support recycled products. The lack of well-defined objectives along with an overarching goal relating to the circular economy hinders its further development. Moreover, lack of funding toward circular economy infrastructure and scant financial incentives in encouraging the use of recycled materials by businesses affects the effectiveness of circular economy (Dhull & Narwal, 2016).

15.8 OPPORTUNITIES

Notable challenges are faced by developing economies in achieving sustainable development. Economic development, environmental collapse, and social injustice are the issues encountered by developing nations in implementing circular economy practices (Khan et al., 2019). Frameworks such as triple bottom line (TBL) and circular economy model have been increasingly important in addressing these complex challenges. The circular economy emphasizes resource regeneration and minimizing waste, while TBL focuses on the need for business having regard to economic, social, and environmental impacts (Tavanti, 2023; Hussain et al., 2023). Through combining the principles of TBL together with the circular economy model, developing countries may address economic, environmental, and social concerns simultaneously (Khan et al., 2021). This merger ensures that there is sustainable economic development, preservation of the environment, and enhancement of social justice (Mirzaei & Shokouhyar, 2023).

15.9 POLICY AND INSTITUTIONAL FRAMEWORK

In India, supportive policies and regulatory frameworks are important to create an enabling environment for the circular economy (Aurisano et al., 2021). By putting into place tax breaks, subsidies, and procurement laws that favor circular practices, India can drive market transformation and support circular business models (Patwa et al., 2021). Such clear rules and regulations play an important role in boosting confidence, thereby encouraging risks and investments in the circular sectors which enhances economic growth and sustainability. In India, the acquisition of certifications and compliance with environmental requirements are important for the promotion of environmental sustainability. Standards such as the ISO 14001:2015 provide a benchmark against which companies can measure and improve their environmental performance (Kafel & Nowicki, 2023). Maintaining environmental standards can promote a culture of environmental accountability, advance sustainable practices in India, and create a framework for everlasting enhancement within the circular economy (Andersen, 2007; Haupt & Hellweg, 2019).

15.10 CASE STUDIES

In India, 12 waste-to-energy plants have been established till now, Delhi is home to the highest number of these plants, that is, 4 (Ministry of New and Renewable Energy, 2017), which produce energy from waste material in a way that reduces pollution and preserves natural resources. The overall amount of solid waste generated in India is 160,038.9 tons/day out of which only 79,956.3 tons/day is processed (MSW Annual Report, 2020–21). The amount of created garbage that is processed is minuscule. It can be clearly seen that there is huge gap between generation of solid waste and its treatment and number of plants for converting waste to energy is not enough. There is more need of such initiatives to treat generated waste in India.

15.11 CONCLUSION

This chapter has explored the synergy between the circular economy and the triple bottom line (TBL) and highlighted its significance in achieving sustainable development. The TBL describes a framework for enterprises to assess their commercial operations and their global impact, while the circular economy acts as a complimentary philosophy encouraging the efficient raw material and waste management. By integrating these frameworks, it is possible to effectively manage the economy, the environment, and the society. The aim of this chapter is to describe how the circular economy is connected with the triple bottom line, thus stressing the need to connect the two for sustainable development. Through the use of this synergy, business organizations and societies can increase economic growth, protect the environment, and promote social justice. This research goes a long way in offering the governments, business organizations and other stakeholders in the circular economy and triple bottom line, optimization of environmental and social resource, that is provision of better services and enhancing social and economic development. In subsequent research, attention should be directed toward the investigation of the circular economy in informal sectors which include waste management, recycling, and repair.

REFERENCES

Aguiñaga, Eduardo, Irene Henriques, Carlos Scheel, and Andrea Scheel. 2018. "Building Resilience: A Self-Sustainable Community Approach to the Triple Bottom Line." *Journal of Cleaner Production* 173: 186–96. https://doi.org/10.1016/j.jclepro.2017.01.094.

Ahmed, Jashim Uddin, Quazi Tafsirul Islam, Asma Ahmed, and Sakib Bin Amin. 2022. "Extending Resource Value-Based Circular Economy Business Model in Emerging Economies: Lessons from India." *Business Perspectives and Research* 11 (2): 309–21. https://doi.org/10.1177/22785337211070363.

Andersen, Mikael S. 2007. "An Introductory Note on the Environmental Economics of the Circular Economy." *Sustainability Science* 2 (1): 133–40. https://doi.org/10.1007/s11625-006-0013-6.

Aurisano, Nicolò, Roland Weber, and Peter Fantke. 2021. "Enabling a Circular Economy for Chemicals in Plastics." *Current Opinion in Green & Sustainable Chemistry*, Elsevier 31. https://doi.org/10.1016/j.cogsc.2021.100513.

Blomsma, Fenna, and Geraldine Brennan. 2017. "The Emergence of Circular Economy: A New Framing around Prolonging Resource Productivity." *Journal of Industrial Ecology* 21 (3): 603–14. https://doi.org/10.1111/jiec.12603.

Dantas, T. E. T., E. D. de-Souza, I. R. Destro, G. Hammes, C. M. T. Rodriguez, and S. R. Soares. 2021. "How the Combination of Circular Economy and Industry 4.0 Can Contribute Towards Achieving the Sustainable Development Goals." *Sustainable Production & Consumption*, Elsevier 26. https://doi.org/10.1016/j.spc.2020.10.005.

De Angelis, R. 2018. "Sustainable Development, Corporate Sustainability and the Circular Economy." In *Business Models in the Circular Economy*, 11–43. https://doi.org/10.1007/978-3-319-75127-6_2.

Dhull, Sunil, and M. S. Narwal. 2016. "Drivers and Barriers in Green Supply Chain Management Adaptation: A State-of-Art Review." *Uncertain Supply Chain Management*, Growing Science 4 (1): 61–76. https://doi.org/10.5267/j.uscm.2015.7.003.

The Economic Times. 2023. "India Recycles Only 30 Per Cent of 3.4 mt Plastic Waste Generated Annually [Report]." *The Economic Times*. https://economictimes.indiatimes.com/news/india/india-recycles-only-30-per-cent-of-3-4-mt-plastic-waste-generated-annually-report/arti-cleshow/96918352.cms?from=mdr.

Ferro, Carlos, Carmen Padin, Nils Høgevold, Göran Svensson, and Juan C. Sosa Varela. 2019. "Validating and Expanding a Framework of a Triple Bottom Line Dominant Logic for Business Sustainability through Time and across Contexts." *Journal of Business & Industrial Marketing* 34 (1): 95–116. https://doi.org/10.1108/JBIM-07-2017-0181.

Geissdoerfer, Martin, Paulo Savaget, Nancy M. P. Bocken, and Erik J. Hultink. 2017. "The Circular Economy—a New Sustainability Paradigm?" *Journal of Cleaner Production*, Elsevier Ltd. 143. https://doi.org/10.1016/j.jclepro.2016.12.048.

Ghisellini, Patrizia, Catia Cialani, and Sergio Ulgiati. 2016. "A Review on Circular Economy: The Expected Transition to a Balanced Interplay of Environmental and Economic Systems." *Journal of Cleaner Production* 114: 11–32. https://doi.org/10.1016/j.jclepro.2015.09.007.

Gurin, J., L. Manley, and A. Ariss. 2015. "Sustainable Development Goals and Open Data, World BankBlogs". https://blogs.worldbank.org/digital-development/sustainable-development-goals-and-open-data.

Hannon, Eric, Clarisse Magnin-Mallez, and Helga Vanthournout. 2016. *The Circular Economy: Moving from Theory to Practice*. McKinsey Center for Business and Environment Special Edition.

Haupt, Melanie, and Stefanie Hellweg. 2019. "Measuring the Environmental Sustainability of a Circular Economy." *Environmental & Sustainability Indicators* 1–2. https://doi.org/10.1016/j.indic.2019.100005.

Hussain, Matloub, Mehmood Khan, and Hussein Saber. 2023. "Thematic Analysis of Circular Economy Practices across Closed-Loop Supply Chains: An Institutional Theory Perspective." *Sustainable Production & Consumption* 40: 122–34. https://doi.org/10.1016/j.spc.2023.06.017.

Indian Express. 2023. "Delhi Ranks # 4 on List of World's Most Polluted Cities." *The Indian Express*. https://indianexpress.com/article/cities/delhi/delhi-ranks-4-on-list-of-worlds-most-pol-luted-cities-8497423/.

Kafel, Piotr, and Paweł Nowicki. 2023. "Circular Economy Implementation Based on ISO 14001 within SME Organization: How to Do it Best?" *Sustainability* 15 (1). https://doi.org/10.3390/su15010496.

Kala, Kaveri, Nomesh B. Bolia, and Sushil. 2022. "Analysis of Informal Waste Management Using System Dynamic Modelling." *Heliyon* 8 (8): e09993. https://doi.org/10.1016/j.heli-yon.2022.e09993.

Khan, Iqra S., Muhammad O. Ahmad, and Jukka Majava. 2021. "Industry 4.0 and Sustainable Development: A Systematic Mapping of Triple Bottom Line, Circular Economy and Sustainable Business Models Perspectives." *Journal of Cleaner Production* 297: 126655. https://doi.org/10.1016/j.jclepro.2021.126655.

Khan, Syed Abdul Rehman, Arshian Sharif, Hêriş Golpîra, and Anil Kumar. 2019. "A Green Ideology in Asian Emerging Economies: From Environmental Policy and Sustainable Development." *Sustainable Development* 27 (6): 1063–75. https://doi.org/10.1002/sd.1958.

Kirchherr, Julian, Denise Reike, and Marko Hekkert. 2017. "Conceptualizing the Circular Economy: An Analysis of 114 Definitions." *Resources, Conservation & Recycling*, Elsevier 127. https://doi.org/10.1016/j.resconrec.2017.09.005.

Korhonen, Jouni, Antero Honkasalo, and Jyri Seppälä. 2018. "Circular Economy: The Concept and its Limitations." *Ecological Economics* 143: 37–46. https://doi.org/10.1016/j.ecolecon.2017.06.041.

Miller, K. 2020. "The Triple Bottom Line: What it Is & Why it's Important, Business Insights Blog". https://online.hbs.edu/blog/post/what-is-the-triple-bottom-line.

Ministry of New and Renewable Energy. 2017. https://pib.gov.in/PressReleaseIframePage.aspx?PRID=1513642.

Mirzaei, Shima, and Sajjad Shokouhyar. 2023. "Applying a Thematic Analysis in Identifying the Role of Circular Economy in Sustainable Supply Chain Practices." *Environment, Development & Sustainability* 25 (5): 4691–722. https://doi.org/10.1007/s10668-022-02217-6.

"MSW Annual Report 2020–21." 2021. https://cpcb.nic.in/uploads/MSW/MSW_AnnualReport_2020-21.pdf.

Mwanza, B. G. 2021. "Introduction to Recycling," 1–13. https://doi.org/10.1007/978-981-16-3627-1_1.

Outlook India. 2023. "India Expected to Face Severe Water Scarcity by 2050: Un Report." https://www.outlookindia.com/national/india-expected-to-face-severe-water-scarcity-by-2050-un-report-news-272364.

Patwa, Nitin, Uthayasankar Sivarajah, Arumugam Seetharaman, Sabyasachi Sarkar, Kausik Maiti, and Kunal Hingorani. 2021. "Towards a Circular Economy: An Emerging Economies Context." *Journal of Business Research* 122: 7235. https://doi.org/10.1016/j.jbusres.2020.05.015.

Tavanti, M. 2023. "Approaches to Managing Sustainability." In *Sustainable Development Goals Series*, 145–170. https://doi.org/10.1007/978-3-031-36907-0_6.

Index

M

N

O

P

R

S

T

U

V

W

£ - 99.99